The Burning House

Appreciation for *The Burning House: A Buddhist Response to the Climate and Ecological Emergency*

One thing is certain: no meaningful systemic change in response to the ecological and climate emergency faced by all planetary life will be possible without a shift in human consciousness. This beautifully crafted, accessible book skilfully weaves Buddhist teachings, a generous account of the author's personal journey and a passionate political and ethical commitment to an empathic path of transformation. Deeply practical, balanced and thoughtful, this book is a trustworthy, insightful companion for the development of the conscious change and enlightened action now so urgently mandated by planetary challenges. – **Anna Grear**, founder of the Global Network for the Study of Human Rights and the Environment (GNHRE); Editor in Chief *Journal of Human Rights and the Environment*

I enjoy how Shantigarbha brings Buddhism and Nonviolent Communication (NVC) together when sharing very alive examples of waking up to our burning house. – **Kirsten Kristensen**, co-founder of LIVKOM, Life-Enriching Communication and Institute of NVC in Denmark

How can Buddhism help us understand and respond to the greatest challenge that humanity has ever faced? The Buddha lived in a very different time and place, but his teachings have important ecological implications for us today. *The Burning House* provides a welcome and very accessible introduction to the relevant teachings and how to embody them in our practice and in our activism. — **David Loy**, author of *Ecodharma: Buddhist Teachings for the Ecological Crisis*

What the world needs now is love in action. Shantigarbha traces a path for us to follow, beginning with attending to our needs. When we use our energy to cultivate our own vitality, we naturally use the abundance we discover in the service of life. Shantigarbha shows how each of us can transform our insight into action, and take affirmative steps to address the suffering wrought from climate change. May all who read Shantigarbha's words be motivated to take action, for the love of all. – **Jim Manske**, CNVC Certified Trainer, author of *Pathways to Nonviolent Communication*

The writer Shantigarbha, as a Buddhist activist, explains the problem through a Buddhist perspective, as a crisis of empathy, connection and community. Through stories from the Buddhist tradition, 'The Burning House', and other stories, he calls for a creative way to face the climate and ecological emergency. As a practitioner of Nonviolent Communication, he guides the reader through practical ways to deal with emotional outbursts that the crisis could provoke. The guided reflection after each chapter draws the reader deeper into the empathic presence. Indeed, a timely book. – **Christlin Rajendram**, S. J., Certified Trainer and Assessor with the International Center for Nonviolent Communication

The ecological crisis is nothing if not a spiritual crisis, a crisis of meaning and direction for our civilization. This book approaches that crisis from a beautifully Buddhist and yet non-denominational perspective. It could help us to awaken.

The Burning House is approachably written and abides by the precautionary principle: moving as we are in a fog, it behoves us to slow down.

Take a pause, and read this book. – **Professor Rupert Read**, author of *This Civilisation Is Finished*, and former student of Thich Nhat Hanh and Joanna Macy

In this book by Shantigarbha, you'll find concrete and accessible ways to engage with the climate and ecological emergency. Sharing his own journey and accumulated wisdom, Shantigarbha offers a truly Buddhist perspective on the emergency – helping us to recognize, deal with our responses to, and find a way out of the 'burning house' that we are in. He evokes what a Buddhist response could look like, from environmental ethics to compassion, from humour to transforming grief into gratitude. The final section of the book then focuses on wisdom-imbued action, with concrete steps on how we might proceed next in our work to put out the fires of this burning house, our precious planet. – **Vajrashura**, Dharma teacher in the Triratna Buddhist Community

The Burning House

A Buddhist Response to the Climate and Ecological Emergency

Shantigarbha

w

Windhorse Publications
38 Newmarket Road
Cambridge CB5 8DT
info@windhorsepublications.com
windhorsepublications.com

Cover design by Katarzyna Manecka
Typesetting and layout by Tarajyoti
Printed by Bell & Bain Ltd, Glasgow

British Library Cataloguing in Publication Data:
A catalogue record for this book is available from the British Library.

ISBN: 978-1-911407-75-1

Contents

About the author

Shantigarbha is a Buddhist mediator and activist. He is an experienced teacher of both Buddhism and Nonviolent Communication (NVC). After studying Classics and Philosophy at Keble College, Oxford, he devoted himself full-time to the Buddha's teachings. Ordained into the Triratna Buddhist Order in 1996, he was given the name Shantigarbha, which means 'Seed or Womb of Peace'. Through his practice of Buddhism he came across NVC and has been sharing it as a certified trainer since 2004.

Within the Triratna Buddhist Community he holds responsibilities as an Order Convenor, a member of the International Council, and supporting restorative practices.

He has trained members of Extinction Rebellion in nonviolence and de-escalation skills.

His first book, *I'll Meet You There: A Practical Guide to Empathy, Mindfulness and Communication*, came out in 2018, published by Windhorse Publications.

For more information about Shantigarbha, Burning House, and Nonviolent Communication courses, visit his website at www.SeedofPeace.org.

Author's acknowledgements

I'm grateful for being alive on this earth, in particular for the love of my partner Gesine and the warmth of the sun on my skin.

I'm also grateful for the support of Dhammamegha, Dhivan, Rowan Tilly, Tejopala, and others who have commented on previous drafts and contributed their ideas to this book. Thank you!

Publisher's acknowledgements

Windhorse Publications wishes to gratefully acknowledge a grant from the Future Dharma Fund and the Triratna European Chairs' Assembly Fund towards the production of this book.

We also wish to acknowledge and thank the individual donors who gave to the book's production via our 'Sponsor-a-book' campaign.

Audio downloads

This book has been produced with accompanying guided meditations and reflections by the author, available as free downloads. They are marked with a and can be streamed directly from the Internet or downloaded in MP3 format. Please go to https://www.windhorsepublications.com/audio-resources/audio-resources-for-tbh/.

Introduction

I want you to act as if the house was on fire.

– Greta Thunberg to the EU Parliament
in Strasbourg, April 2019

When I first heard Greta Thunberg's words, I was immediately reminded of the parable of the burning house from the *White Lotus Sutra*. This is one of the most famous parables of one of the most famous texts of the Buddhist East. It struck me as the perfect metaphor for our times.

In the parable a father comes home to find that his house has caught fire. Then he learns that all of his children are inside the burning house. There isn't enough time to pick them up individually, so he cries out to them in anguish, *The house is on fire. You must get out!* To his dismay, they just carry on playing. In a final flash of inspiration, he hits on a way of getting them to safety. He tells them that there are even more wonderful and exciting toys outside. The children rush out to find these wonderful new toys.

In the parable, the fire represents the dangers of samsara,[1] a life of unawareness.

How do we face up to the existential threat of climate emergency without falling prey to horrified anxiety? How do we do this against the backdrop of the cosmic burning house of samsara? And if we want to get others out of the house, how do we transform our cry of anguish into a cry of inspiration?

This is my witness statement. It's *a* Buddhist response to the climate and ecological emergency. Given the value that Buddhism places on thinking for oneself, it could hardly be *the* Buddhist response. I can hear the fire alarm, and I feel compelled to respond to it now.

I'll take you through an analysis and exploration of what it means to engage with this global crisis here and now. With personal stories, examples, and reflections from Buddhist teachers, I'll take you through the thoughts and feelings you might be experiencing and help give you the freedom to act now.

The chapters follow the Buddhist path from self-development to social change, inspired by the profound principle of non-violence, or love. They also follow my personal journey from ignoring the reports, to coming to terms with the science, to making connections with the Buddhist values of wisdom, compassion, and interconnectedness, to recognizing the social justice implications of the climate and ecological emergency. Along the way, we'll get to strengthen our connection with the earth, bring in some humour, transform our anger, acknowledge our grief, and celebrate our gratitude towards the earth.

I've tried to weave together three themes: using empathy to explore what it means to care about the environment;

looking squarely at the climate and ecological emergency and climate activism; and specifically Buddhist reflections on interconnectedness.

So what could be a Buddhist response to such an emergency? If you have the impression of Buddhists as shaven-headed monks and nuns in the East, or as Westerners sitting in meditation, there might seem no obvious connection with the climate emergency or social activism. And yet Buddhists value mindfulness, wisdom, compassion, and interconnectedness.

Care for the earth seems an obvious channel for acting with compassion in the world. A number of Buddhist teachers have spoken out about the impact of human activities on the environment. So from a Buddhist perspective there is an obvious spiritual dimension to the existential questions that humanity is facing.

All phenomena are interconnected. There are almost eight billion people on this planet, and yet our individual welfare depends on this global community. We now find ourselves in grave danger. We are the only species with the power to destroy the earth as we know it. If we continue as we have done, our civilization will be destroyed. At the same time, we are the only species that has the power to protect the earth.

What would it mean to take on this sense of universal responsibility? We could educate ourselves about the climate and ecological emergency, take the issues seriously, and change our behaviour to protect future generations. We humans created this crisis, so I suggest that it's our responsibility to address it. As the Dalai Lama points out, it's not sufficient to pray for change, it's going to take collective action.[2] What kind of ancestors do we want to be? If we learn to value the future, we can ensure the prosperity of future generations. We can plant trees whose shade we won't live to enjoy.

This is not just a book of ideas about the environment. If you're an eco-activist, I'm guessing that you need support to avoid burnout. This book invites you to explore what resources Buddhism has to offer you to keep going. Could you benefit from further training in nonviolence, love, and wisdom? If you're already following the Buddhist path, I'm guessing that you're looking for ways to make the climate and ecological emergency part of your practice. How do you decide to get involved? For everybody, this book offers ideas and techniques to support an ethical response to the climate emergency and help with the difficult choices and emotions that arise along the way.

The climate emergency is one of a number of profound challenges that we are facing, including loss of habitats and species diversity, reduction in soil quality, and plastic pollution, not to mention systemic differences in ways that people are treated based on the colour of their skin, ethnic group, gender identification, and sexual orientation. Then there are systemic economic inequities and growing resistance to antibiotics. If we successfully tackle the climate emergency, these other challenges will remain. So maybe it's clearer to talk about the climate *and* ecological emergency.

The goal of Buddhism is to wake up to reality. The climate and ecological emergency confronts us with the consequences of our actions and offers us the opportunity for collective awakening. Do we still have time to decarbonize our economies and recarbonize the earth?

At the end of every chapter there are guided reflections to help you engage with your own experience. I suggest that you find a quiet and comfortable place to do these exercises and look for help online if you need advice about sitting posture. It can be useful to have a pen and paper to hand for your thoughts and reflections.

Guided reflection: arriving exercise
Go to https://www.windhorsepublications.com/audio-resources/audio-resources-for-tbh/

- Purpose of this activity: to become more present to your body, your feelings, and your intentions.
- Time: 15 minutes.

1

A crisis of empathy

All living beings are terrified of punishment;
  all fear death.
Making comparison of others with oneself,
  one should neither kill nor cause to kill.
All living beings are terrified of punishment;
  to all, life is dear.
Making comparison of others with oneself,
  one should neither kill nor cause to kill.

– The Buddha[1]

How do we come to terms with the climate and ecological emergency? How do we recognize it as an existential threat? As I search for some kind of reference or comparison, my mind returns to the moment I realized, as a teenager, that I was going to die.

I'd walked out onto the school playing field to take a nap. Lying down on the grass, I daydreamed about a full-sized glider that the school air cadets had built. They'd flown it over the art block and landed it somewhere near to where I was lying. Given what I knew about the air cadets, the whole thing seemed highly improbable, and yet I really wanted to see it rise into the air and glide elegantly over the roof towards me.

Looking up, my attention was caught by a black speck among the billowing white clouds. At first I thought it was a seagull, riding the summer thermals. Then I noticed that it swung back and forth across my field of vision almost as if it was joined to my sight. Scanning left to right, I worked out that the object had to be part of my eye, rather than an object in the sky.

Finally I recognized it as a floater, a few cells joined together in the jelly of my eyeball. When I tried to look at it directly it darted away, like a fish in a stream. I learned how to follow it surreptitiously, catching the occasional, satisfyingly clear view.

Closing my eyes, I saw the orange glow of the sun on the backs of my eyelids. For a moment I was at peace with myself and the world. My mind drifted back to the flight of the glider. Somehow it reminded me of something I'd been thinking about, on and off, for years. Like the floater in my eye, when I looked at it directly it darted away. Could I let it be out of focus for a while, and then swing it into view slowly enough to see it?

I tried this several times, without success. Just as I was about to give up, I realized what I had been trying to bring into my awareness. I realized that *I was going to die*. Just like the improbable glider, my life had a beginning, a middle, and an end.

Within seconds, my mind bounced off this thought. When I tried to return to it by force of will, a deep fear arose. Up until that point I'd always thought that I was immortal, or hadn't thought anything at all. Had I really been missing something so vitally important for so long? The knowledge of my own destruction . . .

After the fear came anger. Why had nobody told me that I was going to die? My parents had told me a lot of things, but they hadn't told me this. It seemed a rather important thing to omit. The experience left me feeling shaky. At the same time, I now felt that I belonged to life. Now I could say, from my guts, that I had as much 'right' to be on the earth as any being, at least for the duration of my flight. I found comfort in the thought that when I died my body would decompose and be assimilated by the earth again.

We simply don't want to know that we're going to die. The thought offends us. Something similar happened to Gautama the Buddha as a young man. There was a moment when he *saw* old age, sickness, and death for the first time. He'd probably seen them before, they just hadn't sunk in. He actually *saw* them when he realized that being a prince didn't give him 'diplomatic immunity'. For the purposes of old age, sickness, and death he was just the same as everyone else. That's quite a leveller.

Of course, he got rather depressed about it, and cut short his pleasure trip to go back to the palace. It's not easy, this business of becoming self-conscious, even if you're a Buddha-to-be. It was only when he saw a fourth sight, of a wandering seeker, that he caught a glimpse of what his life could be, how he could make sense of it all.

The Buddha-to-be had old age, sickness, and death to contend with. That's enough for anybody in one lifetime. Nowadays we also have climate/ecological breakdown. I'd like to suggest that we make it a fifth sight; so we have old age, sickness, death, climate/ecological breakdown, and the wandering seeker. It's something that we need to come to terms with. It's an existential threat that affects all of us, and as with old age, sickness, and death, our mind bounces off it.

I guess that we would need a remarkably happy and peaceful state of mind to entertain the thought that we are collectively harming the earth. That's the earth that we rely on to nourish and sustain us, so it's also a form of self-harm. The mind is repelled by the thought. It's very painful to consider the unfortunate consequences for people and the planet.

So when we read reports on climate and ecological breakdown, it's very tempting to skim the headline and scroll down. It's deeply difficult to read and respond proportionately. What could be proportionate, given the scale of the issues? Giving up your job and becoming a Greta Thunberg-like world saviour?

And yet, how do we keep our sense of humanity, our would-be compassionate nature? How do we come to terms with this fifth sight?

Many of us are experiencing a crisis of empathy and imagination.

As a species, it is difficult to get our heads around the sheer scale of the issue. It is difficult to visualize the data. In ancient Greece people believed that the weather was controlled by the gods. In fact they believed that the weather *was* the activities and movements of the gods. Thunder and lightning were Zeus hurling his thunderbolt, an earthquake was Poseidon striking the ground with his trident, and so on. Humans were powerless in the face of these gods.

Two thousand years later it's still difficult for us to imagine that we mere humans are capable of influencing the weather, that *collectively* we have the power of gods. And yet, from the Industrial Revolution, and especially since 1970, we have the power to change the weather.

One way or another it seems as though we are being forced into a kind of planetary consciousness. It's deeply uncomfortable because our 'circle of influence' has stayed more or less the same, while our 'circle of concern' has expanded more or less infinitely. No wonder that this stimulates profound anxiety for some of us!

Getting our heads around the scale of the climate and ecological emergency means extending our imagination to include all beings, including future generations, ecosystems, and life itself. It means extending our imagination to those who are alarmed about the climate/ecological emergency, those who are concerned, those who are cautious, those who are disengaged and doubtful, and even those who are dismissive.[2] It means empathizing with people (including ourselves) who are continuing with 'business as usual'.

As we shall see later, it even means empathizing with executives in fossil fuel-extracting multinationals. This doesn't mean accepting the status quo. What it means is, humans are always trying to meet needs. Understanding the needs a person is *trying* to meet doesn't mean agreeing with their worldview or the particular way they are going about meeting their needs. Tackling this crisis will involve cooperation on a scale never seen before. No one wants to be told that they are 'bad' or even 'evil' – it stimulates resistance and resentment rather than cooperation. If we want the cooperation of fossil fuel executives, we will need to acknowledge the needs that they are trying to meet, and try to show a different way that is less costly and more fun.

But what do we mean by empathy? Empathy is a respectful understanding of another person's experience. It's an external aspect of mindfulness – a mindful awareness of others and how we communicate with them. Empathy is the reflective, imaginative aspect of compassion. Compassion is the active aspect of empathy. Empathy is the missing link that helps us to develop compassion in a grounded way, towards ourselves and others.

It turns out that the human brain is deeply social. We aren't born loners, reluctantly engaging with others. Humans are essentially group-dwelling, social beings. We are born ready to interact and tune into the people around us. Psychology took a great leap forward when neuroscientists discovered *mirror neurones*, special cells that provide scientific evidence for the proposal that the human brain creates maps of others' minds. Like mirrors, they offer both first-person ('I'm standing here') and third-person ('That's me, that's what I look like from the outside') perspectives.

These neurones fire when someone performs a certain task *and* when they watch another person perform the same task. Previously, action and perception were thought to be controlled by separate areas of the brain. Mirror neurones, or brain areas with mirror properties, bridge these two functions, and allow us to create a map of experience that we can refer to whether it's us performing the action or someone else. Through creating a map of another person's actions and intentions, we can also map their associated feelings and motivations.

But empathy is broader than this; it is also a kind of moral imagination. As such, it's the *basis* for ethics and compassion. Recall the Buddha's words at the start of the chapter, 'All living beings are terrified of punishment; all fear death. Making comparison of others with oneself, one should neither kill nor cause to kill.' The Buddha encourages us to compare ourselves empathically with others. We know *from the inside* that we are terrified of punishment, we fear death, life is dear to us. Looking *outside*, we get the impression that other beings seem to be alive and conscious in the same way as us. Is it the same for these others – are they also terrified of punishment and fearful of death? When we make this comparison, killing another becomes unthinkable.

This is the ethics of empathy. The basis for acting ethically is our capacity to a) connect with the beings around us, and b) imagine the potential impact of our actions. It's not enough just to *feel* a connection, we also need to be able to predict the likely consequences of our actions on other people. This is what distinguishes empathy from *sympathy*, which is feeling the same as the other person, a kind of merging or identification. Empathy recognizes the distinction between self and other, whilst seeing clearly the needs on both sides.

Empathic reflection underpins the first of the Buddha's ethical training principles, or precepts, and forms the cornerstone of nonviolence: 'I undertake the training principle of abstaining from killing.' Or 'With deeds of loving kindness, I purify my body.'

Fear of punishment and death are experiences that we share with all beings. Reflecting on this establishes a common bond of sentience. This is what we have deeply in common.

Life is precious, whether it's *in here* or *out there*. If we're going to identify with anything, we could identify with life itself.

Humans share the capacity for empathy with other mammals such as elephants, apes, dolphins, and whales, and very likely some dogs. However, we seem to be the only species that can empathize with other groups, classes of beings, or species, such as the homeless or polar bears. And we can go further than this. We can extend our imaginative empathy to the natural world, even to inorganic matter or natural events. This is the territory of poetry and art, myth and mysticism.

It's important for us to rediscover the capacity for empathy for the life around us, because this is the basis of ethics. Living in cities, surrounded by technology, it's difficult to maintain a sense of solidarity with life. We tend to find it easier when we are in a forest or other natural place. As my Buddhist teacher, Urgyen Sangharakshita, said, 'The natural world is alive, full of life that resonates with our own lives, and is valuable as life.'[3]

If we see the natural world as alive, rather than as dead, inert matter, raw material for manufacture, or natural capital, we're much more likely to treat it with respect. Philosophically the Buddha's Dharma (teaching) is non-dualistic. This means that it doesn't accept an absolute duality between matter and consciousness. Many thinkers have noted that industrialization and technology emerged in Christian and post-Christian cultures, which accepted the duality between matter and consciousness (or more religiously, worldly flesh and pure soul). Buddhism sees the world non-dualistically, as profoundly alive.

I've been motivated by the ethics of empathy to get involved in environmental action. For others, it might motivate them to engage in science, the study of the natural world. Whether we choose environmental action or science, we need to adopt the long-term attitude of the Iroquois First People and consider the impact of our actions on the seventh generation. We need to be asking ourselves, where are we taking the seventh generation? What will be left for them?[4] Are we harming them by 'taking the not-given'? We'll return to these implications in chapter 4.

Coming back to my experience on that summer's day, I can now see how it brought me alive, and brought everything alive for me. Rather than rejecting the knowledge of my own mortality, I was able to accept it, and this gave me an implicit, gut sense of belonging. It allowed me to empathize with life itself.

Sometimes it helps to get a different perspective. Becoming more aware of your body can be a way of calming the mind and making space for more sustained reflection, including the capacity to empathize more deeply. In the following guided meditation I invite you to lie down on your back with your knees pointing towards the ceiling and take some 'active rest'.

Guided reflection: semi-supine position
Go to https://www.windhorsepublications.com/audio-resources/audio-resources-for-tbh/

- Purpose of this activity: to systematically investigate your bodily experience, in order to calm the mind and create space for reflection and empathy.
- Time: 15 minutes.

2

Crisis? What crisis?

> When you know for yourselves that, 'These qualities are skilful; these qualities are blameless; these qualities are praised by the wise; these qualities, when adopted and carried out, lead to welfare and to happiness' – then you should enter and remain in them.
>
> – The Buddha[1]

In April 2019 my partner Gesine and I travelled from our home in Bristol to attend an Extinction Rebellion march in London. Only days before, the BBC had broadcast David Attenborough's documentary, 'Climate Change: The Facts'. In his familiar, animated way, Attenborough described the ecological toll of human activities. Ironically, he presented the programme from a green field – it could have been the English home counties. The sense of stability and 'nothing could ever go wrong here' from the landscape contrasted strongly with the images of desperation and desolation from elsewhere.

The march was a protest against the recent rate of species extinction. Attenborough's documentary had spelled out the precariousness of many species in the face of climate and ecological change. When Charles Darwin talked about the law of nature being the 'survival of the fittest', he didn't mean 'fittest' in the slang sense of sexually attractive, or in the gym sense of those who can run the fastest or lift the most weights. He meant fittest in the older sense of 'most adapted', most 'fitted' to their ecological niche. Those species that are most finely adapted to their ecological niche survive and thrive. However, this fine tuning can also create a vulnerability. If the ecological niche changes too fast, or even disappears, so does the species. It just can't adapt fast enough, through evolution or changes in behaviour, so it goes extinct.

As we walked at a leisurely pace from our starting point in Parliament Square, I recognized various people from a Buddhist or Nonviolent Communication (NVC) context. I started thinking, I'm sure there's something that I could say about the climate and ecological crisis if I put my mind to it. At Marble Arch, Greta Thunberg addressed the rally.

The stage was the side of a lorry. A woman in her thirties with the Extinction Rebellion hourglass pinned to her skirt said, 'I'm going to bring on two very special people. One of them is Phil, who has been with us from the beginning. He's a veteran activist, arrestee, and grandparent, who puts his body on the line again and again and again, for climate justice. The other is Nobel Prize-nominated, school strike pioneer Greta Thunberg!' The crowd cheered.

An elderly man walked slowly onto the stage and sat down. Greta stood behind him to the side. After a while someone brought a chair for her. In a soft Welsh accent, the man began, 'Hello dear companions, we are in the right place at the right time, and I'm thrilled!' Could it be? No! Yes, it really was Phil Kingston from my Bristol NVC workshops. I recalled the times he'd broken down, overcome by fear of what we are doing to the earth, and how I'd coached him to deliver his message.

Now I watched astounded as he continued, 'WWF [the World Wide Fund for Nature] research tells us that we are using 1.7 earths. . . . The more we are using, the less we are leaving.' No tears, no faltering; just clarity and humour. 'My longing is for the wellbeing and security of my grandchildren and their descendants. And that is never going to leave me, to leave us. We are going to continue. This is our authority. We will build on it and others will follow us. Bless you all!'

Seeing Phil getting his message across brought it home to me. It was the metaphorical straw that broke the camel's back. I was going to *do* something. But what? Phil had been arrested numerous times. I didn't want to go down that route: I am a director of our community-interest training company, Seed of Peace, and a trustee of one of the Triratna Buddhist Community charities. If I get a criminal record, it could affect these responsibilities.

Then there's travel to the USA – I like to offer NVC workshops there and take the opportunity to visit my brother in Indianapolis. Again, having a criminal record could bring an end to this. So I didn't want to be 'arrestable'. Then I remembered my friends who were offering nonviolence training. I thought, 'I could offer my skills as an NVC trainer and mediator!'

In the end the topic of climate change came looking for me. I was invited to speak at the Manchester Buddhist Centre, to launch their Buddhist Action Month. Once I lifted the lid on the topic of the climate emergency, all sorts of intense emotions came out: anger, frustration, anxiety, and fear. I worked hard to create a space where the feelings could be held.

Towards the end of June I led a day workshop on empathy at the Cardiff Buddhist Centre. To my delight, Phil Kingston turned up! I gave him a hug and told him that I was happy to see him. 'I was delighted to see you getting your message across to so many people at Marble Arch!' I said. 'So, you saw that, did you?' he replied in a slow, quiet Welsh tone. 'Well, it was quite a thing. A bit of a miracle really. You see, I was supposed to be interviewing Greta. But when she turned up, she was too stressed. So I had about fifteen minutes to rework my notes into an impromptu talk.' I said, 'You seemed quite relaxed and well-prepared.' He laughed, 'I think that was the intervention of a higher power!'

About the same time as this, I became aware that not everybody shared my conviction that we are in a climate and ecological emergency. I met people who described themselves as climate sceptics. They told me that if I was going to claim that climate change is anthropogenic (i.e. due to human influence) and start disrupting people's lives, then I was morally obliged to do the work to check out the science. I agreed, though I pointed out that I'm not a scientist, so I have obvious limitations here. I told them that I was going on what the IPCC (International Panel on Climate Change) had said. I thought it would be a hard job to argue against them.

The sceptics argued that the IPCC isn't a scientific body but a political body. They believed that it's been corrupted by politics. After these conversations, despite my convictions, I felt a weight of fear on my chest. Was I acting unskilfully?

How could I make sense of their views? As people they seemed sincere, rational, and well-informed. At the same time, as far as I could gather from articles I'd read, there was an overwhelming scientific consensus that greenhouse gases were caused by human activity.

The Dharma has something to say about views, urging us to remember that they are just views. The Dharma encourages us to relate to the *person* rather than the view. These sceptics wanted me to take responsibility for doing my homework and thoroughly investigating the topic before acting on it. I accept that responsibility.

Why are some people sceptical of the science to the point of denial, despite seeming to be rational in other respects? Psychological research suggests that there may be biases at work here.[2] With 'confirmation bias' we tend to select information that supports what we already believe. In the USA and the UK, opinion about climate issues is sharply divided along political lines. Those who identify as right wing see the issue as one of left-wing origin, and reject it on these grounds. This is a reason for not presenting it as a political issue.

Then there may be cognitive biases – a set of inbuilt and largely intuitive mental shortcuts that help us make decisions. These biases help us to apply our previous experience to new information, enabling us to decide what to pay attention to and what to ignore. They are invaluable for dealing with day-to-day decision-making, but can generate serious anomalies when applied to complex issues. In short, humans a) are consistently far more averse to losses than gains, b) are far more sensitive to short-term costs than long-term costs, and c) prioritize certainty over uncertainty.

Climate and ecological issues fall foul of all three of these cognitive biases. Rather than a threat that is concrete, immediate, and indisputable, like an out-of-control truck heading straight for you, climate change can appear abstract, distant, and invisible. It's hard for us to accept that we're playing a game where we've already had the gains and can only look forward to losses. And it's even harder for us to accept short-term costs and reductions in our standard of living, in order to mitigate against much higher, long-term losses sometime in the future. Finally, we crave certainty, and for a long time information about climate change has seemed uncertain and contested. We've tended to score the issue as a draw and to look away.

On top of these inbuilt biases, we now know that the fossil fuel industry spent millions of dollars, over decades, to undermine and minimize the impact of the scientific research on public perceptions and government policies.[3] My own assessment is that I need to talk about this global, existential threat, and communicate how the Dharma can support us in facing it. This is in order to fulfil the solemn declaration I made during my ordination to cherish and protect all beings.

Where to start? If you want things to change, first you need to communicate a sense of urgency.[4] Not panic, not horrified anxiety, but urgency. But is it really a crisis? I decided to start doing my 'homework' and reach out to someone better informed than me. The person I chose was my friend Björn Hassler, who was doing research into climate change when we lived together. Looking him up online, I found that more recently he's been presenting science to the public, amongst other things as the Director of Research at EdTechHub.[5]

I couldn't remember exactly what he'd been researching, so I asked him. He said that he'd been working broadly in the area of climate change, especially how robust the worldwide system of winds in the atmosphere might be. I told him that I just wanted to talk to someone who had more of an insight into climate science than me. He accepted the challenge, so I asked him what he made of the NASA graphs I sent him. Were they really the clincher that climate change is anthropogenic?

He agreed that it was a very convincing presentation. 'Of course it makes sense that increased CO_2 will lead to increased temperatures – that's not a surprise to scientists. The main problem here is the *rapidity* of the change in CO_2 levels. Ecosystems take time to adjust, so there's much more stress for species, including humans.'

I felt relieved and asked him what he would say to people who say that the science isn't settled yet.

He replied that it's important to give appropriate space to different voices. However, if you're driving in the fog, what do you do? Do you speed up or slow down? Some of the climate models predict catastrophic outcomes. This is known as the Precautionary Principle. It comes into play when potential harms have been identified and scientific certainty on the matter is lacking. It emphasizes caution, pausing, and review before leaping into new innovations that may prove disastrous.

What is the evidence? According to a special IPCC report in 2018, greenhouse gas emissions have already driven up average global temperatures by around 1°C, and limiting further warming to 0.5°C would require roughly halving global emissions within the next decade, then cutting them to zero by 2050.[6] Each statement by the IPCC is tagged with the degree of confidence with which they assert it. These statements are tagged 'high confidence'.

Where does their confidence come from? Scientists have been trying to quantify the human influence on climate for decades. In 1999 scientists making use of the UK Met Office's supercomputers and data published a paper asserting that the changes that they were seeing in the surface temperatures were *largely* anthropogenic. Natural factors have contributed very little, particularly since 1950.[7]

Modelling the global climate is notoriously difficult. There's a lot of uncertainty in it, so physicists can't predict directly from first principles. They have to use what's already happened to predict what happens next. They use a combination of models and observations. It's the combination that gives them confidence in their predictions.

The physicists understand, at the level of basic physics, the equations they are trying to solve. The difficulty is that the earth's climate is a turbulent, chaotic system. There are many aspects that they don't understand precisely, so there are many options. They explore these options to see what could be *consistent* with what they observe, rather than using first principles. Hence they create a variety of models to put 'bounds' on a climate forecast. The same applies to weather forecasting – you have to use observations of what the weather is doing now in order to say what it might do next.

So the experimental principle is this: you simulate the system lots of times and throw away the outcomes that don't fit with what you've seen. Then you see what the remainder of the simulations end up doing. For instance, that 1999 forecast is still running and bearing up well.

Since then, computing power has increased dramatically, and it's possible to run a model hundreds of times and see the range of outcomes that it suggests. Initially they used hundreds of thousands of individual PCs, each making predictions based on the parameters that were fed in. Then they asked to what extent these predictions overlap with observations of the climate over the last fifty to a hundred years.

Does this sound to you like 'good' science? What I recall from my Philosophy of Science classes at university is that science boils down to creating a hypothesis and testing it experimentally. You select a hypothesis to investigate by assessing its explanatory force and predictive power. That is, you give more weight to the hypothesis that most accurately explains past and current events and predicts the future.

In this case scientists have been running and testing computer models, which could be said to contain multiple hypotheses. However, the experimental procedure, known as falsifiability, is the same.[8] I've seen no credible evidence of the kind of widespread scientific conspiracy that would have to be going on to discredit this approach and therefore the findings of the IPCC.

In 2015 nations pledged to limit global warming to 1.5–2°C above pre-industrial levels as part of the Paris climate accord. But those governments are largely failing to meet their commitments, and emissions of carbon dioxide and other greenhouse gases continue to rise. Current national commitments have put the world on a path of around 2.6°C of warming compared with pre-industrial times.[9]

As the earth warms, various feedback loops kick in. The Arctic, for instance, is warming two to three times faster than the rest of the globe. As the snow and ice melt, less of the sun's energy is reflected back into space (the Albedo effect), leading to further warming. Melting permafrost exposes vast quantities of microbes, hungry to eat the organic matter previously trapped alongside them. This digestion releases vast quantities of CO_2 and methane, which is thirty times more potent a greenhouse gas than CO_2.

Is it an emergency? On geological timescales it's unquestionably an emergency. Compared to similar changes in the past that have had a very traumatic impact on life on earth, things are changing in the geological equivalent of a twinkling of an eye. The last time CO_2 concentrations were this high – passing 400ppm (parts per million) in 2013 – was in the Pliocene, three to five million years ago. Temperatures then were 2–3°C warmer and sea levels 10–20 metres higher than they are today.[10] CO_2 levels are projected to reach 450ppm by the year 2040.

In our fossil-fuelled growth spurt we've taken billions of tonnes of carbon that have been buried for millions of years and added them to the regular carbon cycle. It's much more than can be absorbed by the biosphere, so the percentage of carbon in the atmosphere has been growing exponentially. As those NASA graphs showed, there is a nearly linear relationship between these cumulative emissions and the temperature rise.

This allows the IPCC to work out a carbon budget: how many more billion tonnes of carbon can we put into the atmosphere and still have a reasonable chance of restricting temperature increases to 1.5°C by the end of the century? To have a 50 per cent chance of limiting it to 1.5°C, our carbon budget is only 395 gigatonnes, or 235 gigatonnes for a 66 per cent probability. To put this in perspective, 2019 emissions were around 50 gigatonnes, so with business as usual we will use up the remaining budget within approximately eight years, or five years for a 66 per cent probability.

Average global atmospheric temperatures have risen and will continue to rise. It turns out that our climate, and our civilization, is extremely sensitive to these rises. For example, most of our cities are located on the coast. In terms of quality of life on earth, we're paying for dumping the carbon into the atmosphere now, and will continue paying for it for centuries to come. Even if we stop putting carbon into the atmosphere, global temperatures will continue to rise. To me this sounds like a fire alarm going off. What does it sound like to you?

As if a climate emergency wasn't enough, we are also in an *ecological* emergency. A 2019 report by the biodiversity equivalent of the IPCC warned that nature is declining globally at rates unprecedented in human history. The rate of species extinctions is accelerating, with grave impacts on people around the world now likely. The report found that around one million animal and plant species are now threatened with extinction, many within decades, more than ever before in human history. The health of ecosystems on which we and all other species depend is deteriorating more rapidly than ever. We are eroding the very foundations of our economies, livelihoods, food security, health, and quality of life worldwide.[11] Another fire alarm.

But how do you make up your mind? The Buddha said, use your own experience and the testimony of the wise. In relation to climate science, the wise are the scientific experts. The evidence is there. The question is, do we care? Do we care that islands such as Kiribati are now under water? Can we empathize with people who are impacted by our decisions? Are we willing to take that into account in decisions we make? Where in our equation does it come?

It's difficult for us to imagine things that will impact us, especially over long periods of time. What can we imagine here in the UK – flooding maybe? Migration is more likely to be an issue: populations on the move, wanting a more secure life in Europe – and why wouldn't they? Perhaps crop failure due to climate change. Westernized countries, which have more resources, will be able to adapt more easily; subsistence farmers in sub-Saharan Africa will find it difficult to adapt.

The Buddha once made a visit to the Kalamas, a community faced with a very modern dilemma. Let's call it fake news! They told the Buddha that various teachers and sages came to their town to teach. They glorified their own teachings, and disparaged, reviled, and showed contempt for other teachers' doctrines. So which of these teachers and sages were speaking the truth, and which ones were lying?

The Buddha began, very wisely, by empathizing with their dilemma. 'Of course you are uncertain, Kalamas. Of course you are in doubt. Where there are reasons for doubt, uncertainty arises.'

Then he advised the Kalamas not to take the following common sources of information as *certain knowledge* without further examination:

- Reports, simply because the source seems reliable.
- Legends, simply because they are held to be meaningful.

- Traditions, simply because this is how we've always done things.
- Scripture, simply because it is taken as the word of God.
- Logical conjecture, simply because it seems logical.
- Inference, simply because it seems logical – the premises may be inaccurate.
- Analogies – the analogy itself may be inaccurate.
- Agreement through pondering views – just because you've taken your time to think things through, it doesn't mean that your conclusions are accurate.
- Probability – just because something seems probable, it doesn't mean that it's accurate.
- By the thought, 'This contemplative is our teacher' – simply relying on the authority of your teacher.

This seems like a pretty thorough list! When they asked him what they *could* rely on, he offered these two criteria, to be taken in conjunction with each other:

1. 'When you know for yourselves that these qualities, when adopted and carried out, lead to welfare and to happiness, then you should enter and remain in them.' In other words, any view or belief must be tested by the results it yields when put into practice over time.
2. 'When you know for yourselves that these qualities are praised by the wise.' In other words, to guard against the possibility of any bias or limitations in one's understanding of those results, they must further be checked against the experience of people you regard as wise.[12]

How can we apply these criteria to the climate and ecological emergency? What can you vouch for in your own experience?

Have you tested your views or beliefs to see what results they yield when put into practice? Do they lead to welfare and happiness for you and others?

Now take a moment to reflect on who you regard as wise. In my case, I regard the scientists who work on climate issues and publish their research in peer-reviewed journals as wise. Bring to mind someone that you regard as wise. What have they said, if anything, about the climate and ecological emergency?

The alarm is going off. Can you hear it? Are you willing to hear it? In terms of the parable of the burning house, will you be the adult who notices the danger or one of the children, distracted by toys and ignoring the smoke and the sound of the alarm?

Many individuals have heard this fire alarm, so why haven't we been able to hear it collectively? Why haven't we been able to take the appropriate steps to mitigate the existential risks? Action by Westernized democracies has been limited to setting CO_2 reduction targets twenty to thirty years in the future. To my mind this is kicking the can down the road. The trouble is that the road might well be scorched or under water by the time we get to it.

It's easy to become fatalistic and believe that there's nothing to be done. The forces at play are too great; the powers that be can't and won't adapt. However, we always have a choice. Once, in my jet-setting days, I got into a panic as I was packing my bag for a foreign workshop trip. I was still frantically packing when the taxi arrived to take me to the airport. Ten minutes later, as I got into the taxi, I said to the driver, 'Phew! That was hard work. I don't think I've ever been so stressed about packing to go abroad.' He fixed me in the mirror and uttered these words, which I'll never forget, 'You could have been ready yesterday, if you'd wanted to!' I was stunned into silence.

My sense is that it's overwhelming for us to hear about ice melting in the Arctic, reduced numbers of polar bears, and so on. Collectively, we don't seem to want to hear about it. I guess that we want to do our jobs and take care of our families. I guess that we want to enjoy our lives. Perhaps we're just not very good at dealing with existential threats.

Governments both reflect and perpetuate the attitudes of the people that they govern. Around the world they have repeatedly said that there's no way we can shut everything down in order to lower emissions, slow climate change, and protect the environment.

However, the Covid-19 pandemic showed us that governments will take drastic measures if they have popular support. In a surprising outbreak of collectivism and compassion, governments prioritized saving lives over economic considerations. Perhaps it's not that surprising: shared risks create a sense of community. People willingly accept shouldering their part of a collective burden, like the hardships of lockdown, when they share a common purpose and are rewarded with a greater sense of social belonging.

And therein lies much of our hope.

Trying to come to terms with the climate and ecological emergency can be deeply challenging. It can be hard to come to our senses in the context of such an existential threat. Perhaps this is a moment for self-connection, for coming to our senses in quite a literal way. I invite you to join me in a guided arriving exercise, in which I will lead you to a deeper awareness of your body, your feelings, and your underlying needs and values.

Guided reflection: coming to our senses
Go to https://www.windhorsepublications.com/audio-resources/audio-resources-for-tbh/

- Purpose of this activity: to become present to your body, feelings, and needs.
- Time: 15 minutes.

3

Touching the earth

This earth supports all beings;
She is impartial and unbiased toward all,
whether moving or still.
She is my witness that I speak no lies;
So may she bear my witness.

– Siddhartha, the future Buddha, on
the eve of his Enlightenment[1]

The root of the climate and ecological crisis is how we human beings relate to the environment. What can we learn from the Buddha's example? What is a meditator's perspective?

The Buddha's quest and Enlightenment were characterized by liberation from the 'round' of birth and death. We can get an insight into how the Buddha related to the natural world by going back to the eve of his Enlightenment, when he was still Siddhartha Gautama, sitting under the bodhi tree.

What did the Buddha-to-be look like? There was no tradition of portraiture in the modern sense. For the first few hundred years after the Buddha's life, his image was regarded as too precious to depict in human form. Instead, artists represented him by a sign such as a pair of footprints, an empty seat, or a space beneath a royal parasol. So every culture and generation needs to reimagine the Buddha's appearance.

We have some clues: he was born into a warrior prince (kshatriya) family, the only son of a local raja (elected tribal king), in north-east India, just inside the modern Nepalese border. This suggests that he may have looked like modern Nepalese people from the foothills of the Himalayas. After he had gone forth from his royal life to seek the truth, he walked south through another kingdom. The local king saw him from his palace, and the first thing he noticed was Siddhartha's demeanour – handsome, stately, and pure.[2] In fact the king was so struck by Siddhartha's appearance that he offered him a position as a warrior prince in his army.

Sitting under the bodhi tree at Bodh Gaya, Siddhartha must have retained his regal appearance, tempered by the hardships of six years of extreme ascetic practices. His limbs were relaxed in meditation and communicated a sense of groundedness and oneness with his surroundings. Finally, his expression was unbelievably radiant and peaceful.

As he's sitting there, Siddhartha comes to the attention of Mara, the somewhat pantomime devil of the Buddhist scriptures. Mara realizes that if Siddhartha becomes awakened, he will awaken billions of beings and empty Mara's realms. So he summons his sons and daughters and goes with a great army to defeat the Buddha-to-be. As well as trying brute force and seduction, he challenges Siddhartha's right to sit on the seat of Enlightenment. Mara asks, 'On what grounds could you ever attain liberation?'

Siddhartha replies that he has performed trillions of unstinting acts of giving. He has even sacrificed his own body in order to liberate sentient beings. Mara challenges him, declaring that he has no witnesses to his acts. Siddhartha replies, 'The earth here is my witness.'[3] Without fear, he lets his right hand slide down and taps the earth. The earth shakes in six different ways, and the earth goddess, named Sthāvarā, breaks through the surface near Siddhartha. She bows down to him and bears witness to his altruism over many lifetimes.

So Siddhartha claims his right to sit on the seat of Enlightenment with a gesture of groundedness, of touching the earth. He calls the earth to witness his generosity over countless lifetimes. We have accounts of some of these previous lives in the Jatakas, or birth stories. Many are tales of animals sacrificing themselves for other animals. It turns out that the ones who sacrificed themselves were the Buddha in a former life. It's this profound generosity that gives Siddhartha the right to sit under the bodhi tree, on the seat of Enlightenment.[4]

Mara is defeated by this gesture and by the witness of the earth goddess. Surprising as it may seem, the Buddha-to-be achieves this victory with a gesture of groundedness.

The episode of Mara's attack and defeat has a parallel in the Bible. During Jesus' forty days in the desert, Satan comes to tempt (test, challenge) him. Instead of touching the earth, Jesus quotes scripture at Satan. When Satan challenges Jesus to perform a miracle by turning stones into bread, Jesus says, 'Man may not live by bread alone', turning the story into an allegory of worldly and divine meaning.

Establishing meaning by touching the earth is entirely in keeping with the rest of what we know about the Buddha.

We saw in his advice to the Kalamas that he doesn't rely on scripture as a way of establishing meaning. Instead, he relies on what we might call 'embodied cognition'. Touching the earth is a gesture of embodied cognition. As modern psychology testifies, we learn and remember with our bodies. Our bodies remember the lessons of our lives. It is a direct way of knowing and seeing, intimately connected to our surroundings and the natural world.

The Buddha had a particularly close connection to trees. He was born in a forest, lived in the forest after going forth, and gained Enlightenment under the bodhi (peepul or *Ficus religiosa*) tree. Afterwards he spent seven weeks sitting within sight of that tree, absorbing the profound impact of the experience.

In the second week, the newly awakened Buddha looked back at the bodhi tree, feeling deep gratitude for how the tree had sheltered him. This reminds us that gratitude is an Enlightened quality – knowing that something has been done for our benefit. Veneration of the bodhi tree continues to this day in Bodh Gaya. Recent archaeological research at Lumbini, the Buddha's birthplace, suggests that the shrine originally protected a tree.

In the time of the Buddha, people believed that trees were inhabited by spirits. Not just trees, but also hills, woods, and groves could become shrines.[5] This form of animism (the belief that nature is alive) is echoed in the myths of ancient Greece, Rome, and other cultures. The Buddhist tradition used Jataka stories about trees to illustrate values of care, protection, and ecological awareness.[6]

In one story about the Buddha's former life as a sal tree in the Himalayas he exhorts other trees to live in the safety of the forest. A great storm sweeps through the region and the trees in the forest survive together, whereas solitary trees sustain damage or destruction, showing that there is strength in unity.

In another story a beautiful wishing tree is going to be cut down to repair the main pillar of the king's house. The king's carpenters bring sacrifices to the tree and let the tree know that they will be back the following day to cut it down. The tree bursts into tears and all its friends gather round to find out what's the matter. None of them can think of a way to stop the carpenters. Finally the future Buddha, who is living beside the wishing tree as a clump of grass, transforms into a chameleon and turns the wishing tree's bark blotchy. When the carpenters arrive they declare that the tree is rotten and has holes. The wishing tree is saved! This fable speaks to the value of ingenuity in preserving ecosystems.

Finally a lion and a tiger live in a forest. While they are there, humans dare not enter. However, one of the tree spirits becomes disgusted with the stench of animal carcasses left by the lion and the tiger, so the tree spirit scares them off. Once the humans realize that the lion and the tiger are no longer in the forest, they enter it, clear it, and cultivate the land. The moral of the story is to look at the ecosystem *as a whole* before intervening to improve it.

This story brings us to the ecological situation at the time of the Buddha. The landscape of northern India in the fifth century BCE was undergoing changes. There were still large tracts of virgin forest, home to rhinos, elephants, tigers, and large snakes. At the same time, urban centres were growing and forest was being cleared for agriculture. The Buddha and his wandering disciples renounced their home lives and lived in the forest, or on the edges of villages and towns, at the intersection between civilization and wilderness. They allowed very few comforts to come between them and the natural world.

It wasn't only trees that the Buddha held in reverence. He revered the whole of life. He believed that there was continuity of life between humans and animals,[7] perhaps even trees and plants, as in the Jataka tales. Practising meditation, the Buddha and his followers naturally became sensitive to their surroundings.

The Buddha's reverence manifested in care and protection for the environment and natural resources. At one time he paid a visit to three of his disciples who were living together as a community in a secluded grove. These three practitioners, known collectively as the Anuruddhas, followed an intensive programme of meditation and Dharma study. As a key part of their practice they maintained harmony with each other, regarding each other 'with kindly eyes, blending together like milk and water'.[8]

They also lived in harmony with the natural world. As they describe their daily habits and agreements to the Buddha, one detail stands out. In the mornings they go on their alms round to the local village to collect food. The last one to return from the village eats any leftovers, if he wishes, or throws them away where there is no greenery, or drops them into water where there is no life. With this practice they avoid polluting water sources for their own use and for the animals, birds, reptiles, plants, mosses, and even bacteria that also depend on them. We forget that ancient people had an acute ecological awareness, because their survival, and the survival of their environment, depended on it.

The Buddha extended his protection to plants in general. He laid down a monastic rule that damaging a living plant was an offence, which monks were to confess in the monthly confession ritual.[9] The later commentator Buddhaghosa explained how this rule came about. It started when a certain monk chopped down a tree to build a hut for himself. The spirit who lived in the tree came forward and complained to the Buddha that the monk had cut off his child's arm. The Buddha's response was twofold: he gave the tree spirit another tree to live in and he laid down the rule that damaging a living plant was a confessable offence.[10]

The Buddha made other monastic rules, such as not travelling in the rainy season, in order to avoid killing the worms and insects that come to the surface during this time; not wearing wooden clogs (for the same reasons); treating wild animals with kindness; revering large trees; and protecting trees and open lands.[11] He instructed his monks to filter all their drinking water, in order to avoid drinking even the smallest of creatures.

Modern technology has made our lives easier and more comfortable in many ways. It has also made communication between us easier. However, left to our own devices(!) we can binge on news, TV, politics, celebrity culture, social media, movies, constant messaging, screens, screens, screens. We are faced with a deluge of information, some of which is comforting and connecting, some of it distressing.

Spending hours each day in front of a screen can have an ungrounding effect. We may experience tension in our neck and head, and our centre of consciousness will tend to rise up from our chest to our head. Just as the Buddha-to-be was able to ground himself in the face of Mara's attack by reaching down and touching the earth, so we can ground ourselves through connection with the natural world.

Even 2,500 years ago the Buddha exhorted his monks and nuns to follow his example and get outdoors. In one of the key early meditation texts, the Buddha suggests that they go to the wilderness, to the shade of a tree, or to an empty building.[12] Forests provide natural, undisturbed, quiet, and peaceful places for meditation. They are also beautiful. In one discourse Sariputta, a chief disciple of the Buddha, declares, 'The forest of sal trees at Gosingha is highly delightful, the night is brightly moonlit, and the sal trees are spreading a delicate fragrance, as if from divine flowers.'[13]

It's worth noticing that the Buddha and his disciples followed a lunar calendar. They gathered at night around the time of the full moon each month to chant, meditate, and confess. This was partly practical – the days could be extremely hot, and the light of the full moon enabled them to find their way along tracks and roads. However, ancient people, including Buddhists, were also acutely aware of the power of the full moon to give added intensity to any kind of spiritual practice. The Buddha himself reached Enlightenment during the night of the full moon in May/June.

Living largely in the open air, sensitive to the cycles of the moon and the seasons, the Buddha and his followers were able to remain grounded. For us, living in the era of artificial light and gadgets, surrounded by buildings and people, getting our food from shops rather than directly from the soil, it's easy to lose touch with the earth.

Another disciple, Maha Kassapa, said to be the founder of the Zen tradition, gives us one of the first nature poems:

Strung with garlands of flowering vines,
This patch of earth delights the mind;
The lovely calls of elephants sound –
These rocky crags do please me so!

The shimmering hue of darkening clouds,
Cool waters in pure streams flowing;
Enveloped by Indra's ladybugs –
These rocky crags do please me so!

Like the lofty peaks of looming clouds,
Like the most refined of palaces;
The lovely calls of tuskers sound –
These rocky crags do please me so!

The lovely ground is rained upon,
The hills are full of holy seers;
Resounding with the cry of peacocks –
These rocky crags do please me so!

Being clothed in flaxen flowers,
As the sky is covered in clouds;
Strewn with flocks of various birds –
These rocky crags do please me so!

Not occupied by village folk,
But visited by herds of deer;
Strewn with flocks of various birds –
These rocky crags do please me so!

With clear waters and broad boulders,
Holding troops of monkey and deer;
Covered with moist carpets of moss –
These rocky crags do please me so![14]

On his farewell teaching tour, aged eighty, the Buddha appreciated the places he had known as he said goodbye to them. He remarked on the pleasantness of Vaishali and its shrines. He enjoyed the town of Rajagriha and its majestic Vulture's Peak. When he left Vaishali for the last time, he looked back with an 'elephant look', turning his whole body to observe the scene.[15]

Whether we live in the mountains or not, sensing the earth element can give us grounding. From a meditator's perspective, we can sense the earth element *internally* as the quality of solidity in our body. We can find this quality in our bones, muscles, and sinews. We can also sense the earth element *externally* by feeling the solid ground beneath us as we sit and imagining how far the earth stretches into the distance.[16]

Again from a meditator's perspective, we can reflect on how we *depend* on the natural world for our survival. The human body depends on nourishment from the earth; it is not complete and indivisible. We depend on appropriate living conditions. The existence of the body depends on the existence of the earth.[17]

Ancient people of both East and West saw physical form as having four qualities: earth, water, air, and fire. In Buddhist thought these four are not regarded as substances in their own right, but rather as tendencies or qualities: earth represents the quality of solidity, water the quality of fluidity, air the quality of lightness and expansiveness, and fire the quality of cold/heat and radiation.

Our bodies are in a constant exchange with these elements. We exchange with the earth element in the form of food, with the air element in form of oxygen, with the fire element in the form of warmth, and with the water element in the form of drinks. This exchange is our connection to life, our sense of aliveness.

Take breathing, for example. If it stops, we stop. We breathe in oxygen that has been exhaled by trees, plants, and plankton. In turn they breathe in our CO_2 and transform it back into oxygen. On geological timescales we can only survive as mammals because of the millions of years that plant life has been taking CO_2 out of the atmosphere and turning it into oxygen plus carbon, in the form of soil, rock, coal, and oil.

We can reflect on the four elements as an example of the Buddha's insight that things arise dependent upon conditions: 'This being, that becomes.' When conditions are in place, things arise. When those conditions cease, those things cease.[18] This kind of reflection can help to undermine our existential belief that our 'self' is somehow fixed and indivisible. In particular we can use it to reflect on impermanence. If even the earth is impermanent, how much more so is this human body?

We all depend on suitable living conditions, both now and in the future. Humans need nature to thrive, not the other way round. Without suitable living conditions humans won't be able to practise and attain Enlightenment. Whatever we do to the earth, to nature, ultimately we do to ourselves. So protecting the natural world is in the general interests of humanity. For Buddhists it has a particular value in supporting the possibility of Enlightenment. This is a key way of establishing environmental ethics, which is the subject of the next chapter.

Facing the climate and ecological crisis can be a path to awakening, if we approach it with love and awareness. The Buddha was keenly aware that even the most substantial, enduring objects in the world, like mountains and oceans, are impermanent. If these vast objects are impermanent, how much more so are we.

Generally speaking, Buddhism is not an 'end of the world is nigh' tradition. The Buddha taught that samsara is endless. Only by seeing through the cycle of birth and death can we end it. He established a community of practice, which was not something he would have done if he believed, like Jesus did, that a cataclysmic event like the coming of the kingdom of God would end all normal life.

However, in one discourse he asks his followers to imagine the end of the world.[19] What would happen if the rains stopped indefinitely? In time all the trees, grains, and medicinal herbs would dry up and disappear. He reminds his audience that all phenomena are impermanent, subject to change, not to be relied upon, something to seek freedom from.

Then what would happen if a second sun, as strong as our present sun, appeared in the sky? Streams and rivulets would dry up. And a third sun? The great rivers would dry up. A fourth? The sources of all the great rivers would dry up. A fifth: the great ocean would recede and dry up. A sixth: the earth with its mountains would emit a mass of smoke. And finally, a seventh sun: the earth with its mountains would ignite into a single mass of flame. It would burn and be destroyed, without even ash remaining.

Of course, we are modern people, influenced by scientific ideas. I guess that your geography teacher didn't mention seven suns. We have a different idea of what will happen at the end of our solar system. Apparently, in about five billion years our sun will become a red giant, engulf Mercury and Venus, and probably the earth. Then it will shrink to become a white dwarf, and no longer produce energy by fusion. Finally, trillions of years later, it will become a black dwarf.

Most of us can contemplate this sequence of events with relative equanimity because it is so far in the future. On these timescales we can recognize that all phenomena are impermanent and insubstantial. It becomes harder when we try to imagine the end of the weather systems that we grew up with, as well as flooding, mass migration, and a loss of the benefits of 'civilization' such as interruptions to the food supply, *in our lifetimes*.

Could we learn to accept this possibility, without falling prey to despondency or fatalism? Could it be a spur to practice and compassionate activity? What would we need to understand, accept, and let go of, in order to do this?

It's very important for us to sense the ground on which we sit, to allow the earth to take our weight. We may feel such a sense of displacement, guilt, and anxiety around our collective impact on the earth that it's difficult even to do this. Apart from giving you a chance to relax and centre yourself, it will also give you a sense of belonging to the earth. I invite you to join me in a guided reflection to experience this sense of groundedness and belonging for yourself.

Guided reflection: four element practice
Go to https://www.windhorsepublications.com/audio-resources/audio-resources-for-tbh/

- Purpose of this activity: to allow the earth to take our weight. To reflect that our body relies on earth, air, fire, and water and that we are in a constant exchange with these elements.
- Time: 20 minutes.

4

Environmental ethics

An apology

We live in the Age of Apologies. Here is an apology that is much more meaningful than many being made today.

[Hu]mankind owes a profound apology:

To the Birds, for having polluted the air through which they fly,
To the Ape and the Tiger, for having destroyed the forests in which they live,
To the Deer and the Bison, for ruthlessly hunting them almost to extinction,
To the Rivers and Streams, for poisoning them with chemicals,
To the Earth itself, for greedily pillaging its riches of silver and gold,
To the Ocean, for slaughtering the greatest of her children, the Whale, 'for scientific purposes',
To the Mountain Peaks, for defiling their virgin snows with our trash,
To the Moon, for rudely invading her sacred space,

To the Stars, for obscuring their brightness with the smoke of our cities,
To the Sun, for not gratefully acknowledging our dependence on his bounty,
To the truly great Men and Women of the past, for not honouring their memory as we should and for not walking in their footsteps.

– Sangharakshita, July 2009[1]

What specific support does the Buddhist tradition offer to us in the climate and ecological emergency? What do we mean by a Buddhist environmental ethic?

At its most basic level Buddhism offers a path of development to realizing our potential as human beings. This path has two phases: a path of vision and a path of transformation. The vision may be quite brief. We might get a glimpse of how things could be, or of our potential, through meditation, a spontaneous epiphany, or other life-changing experiences. My glider moment from chapter 1 is one example. However, according to Buddhism, this kind of experience is not the end of the matter. Then comes the slow, sometimes challenging path of transformation in accordance with that glimpse of reality.

The Buddha broke the path of transformation down into the three great phases of ethics, meditation, and wisdom (insight). It's a progressive path: becoming more ethical in our behaviour supports deeper meditation, which in turn supports greater insight. So when we're considering Buddhist environmental ethics, we need to bear in mind that ethics comes as the great first phase in the path of transformation.

The guided reflections at the end of the chapters in this book are designed to support the second and third phases of meditation and wisdom (insight).

But what of vision? What does the Buddha's teaching have to offer as a vision that is relevant to this emergency? We can find it in the Mahayana vision of interdependence. The Mahayana, or 'Great Vehicle', was a later Buddhist movement that re-emphasized the importance of compassionate, altruistic activity. It universalized the potential for Enlightenment to include all beings, times, and places. It also expanded the earlier (Hinayana or 'Lesser Vehicle') teaching of conditionality or dependent arising to a vision of interdependence.

In the earlier teaching on conditionality, the Buddha trained his followers to undermine any fixed sense of self by investigating the multiple conditions upon which that sense of self arises. In the Mahayana sutras the Buddha restates this as an *ecological* understanding. Nature, and all phenomena, are seen as relational because they are dependent on multiple conditions.

The *Avatamsaka Sutra*, which emerged in the Hua Yen school of Buddhism in seventh-century China, refers to the image of Indra's net to illustrate the infinite complexity of these conditions.[2] In Indian mythology Indra was the king of the gods, rather like Zeus in the ancient Greek pantheon. He owned a magnificent net, so large that it stretched indefinitely in all directions. In accordance with Indra's opulent tastes, the maker had tied a single glittering jewel into each one of the net's infinite nodes. Since the net was infinite in dimension, the jewels were infinite in number. This would surely have been a marvellous sight to behold.

However, if we take a closer look at any one of these jewels, we can see something even more marvellous. Each of these infinite jewels emitted light, and in its polished surface we can see reflected all the other jewels in the net, infinite in number. And each of the jewels reflected in the facets of this jewel is reflecting all the other jewels, so that the process of reflection is itself infinite.

The image harks back to Buddhism's origins in India, but it also looks forward to the World Wide Web, a distributed network in which no node or nodes can be said to be central. It symbolizes a cosmos in which there is an infinitely repeated relationship among all the constituents. The relationship is characterized as one of simultaneous mutual intercausality, or dependent arising.[3]

We can apply this metaphor directly to our current climate and ecological emergency. If any of the jewels become cloudy, they reflect the other jewels less clearly. If any ecological niche in our environment becomes toxic or polluted, it affects all the other niches. A loss of species or habitat in one place affects the rest of the environment. Likewise, if rivers are cleaned and wetlands restored, life across the environment is enhanced.[4] This net of conditions accounts not only for physical and biological factors, but also for human intentions. In fact, human intentions are critical in determining what happens to the net.

The Mahayana tradition presents the Bodhisattva as the ideal Buddhist. A Bodhisattva is a being who dedicates themselves to Enlightenment for the sake of all beings. Enlightenment becomes *waking up to reality for the benefit of all beings*. In this context, it's not surprising that an aspiring

Buddha would respond to an existential threat to all life with a sense of urgency. The Tibetan Buddhist tradition goes further, encouraging us to treat all sentient beings as if they had been our loving mother in one or more of our innumerable previous lives. How about you – does this inspire you?

Now that we've seen some of this vision of reality, let's go back to the earliest Buddhist texts to establish the basis of Buddhist ethics. I've already mentioned that human intention or motivation is key. This is one of the Buddha's key insights. Acting ethically isn't simply about following rules or commandments. He realized that it is the state of mind with which you act that is most important, not the precise performance of actions. Volition is the active ingredient that determines the consequences of an action, whether happier or unhappier.[5]

He formulated his understanding in the first two verses of the *Dhammapada*, perhaps the most famous early Buddhist text:

> 1. Experiences are preceded by mind, led by mind, and produced by mind. If one speaks or acts with an impure mind, suffering follows even as the cartwheel follows the hoof of the ox (drawing the cart).
>
> 2. Experiences are preceded by mind, led by mind, and produced by mind. If one speaks or acts with a pure mind, happiness follows like a shadow that never departs.[6]

We need to pay attention to our states of mind because the states of mind with which we act will largely determine the outcome. In other words it's the *how*, rather than the *what*, that will determine whether our actions lead to happier states for ourselves and others, or the opposite. This is the ethics of intention, or more precisely, of volition. So when we're acting to address the climate and ecological emergency, we need to bear this in mind. The more wholesome our volitions, the more likely our actions will contribute to the welfare of all.

As the Buddha explained in his famous Fire Sermon, everything is burning. We're back to the burning house! The eye, its objects, and associated feelings are burning. The same is true for the other senses: the ear, nose, tongue, body, and mind (the Buddha counted the mind as a sixth sense). Our senses are on fire with greed, hatred, and ignorance. They are burning with birth, aging, death, sorrow, lamentation, pain, grief, and despair. In other words, greed, hatred, and ignorance drive us to experience a mass of suffering. It's only through becoming disenchanted with the senses and their objects that a follower of the Buddha gains equanimity and release.[7]

At first glance, this could give the impression that Buddhist ethics is about right and wrong. However, the Buddha didn't primarily talk in these terms. He usually categorized volitions as wholesome or unwholesome, skilful or unskilful. In the Fire Sermon actions based on greed, hatred, and ignorance tend to lead to suffering for oneself and others. By contrast, actions based on generosity, love, and awareness lead to happier states. This is the ethics of intention or volition.

However, we can always act *more* skilfully or *less* skilfully. Crucially this depends on awareness and imagination. We need to be aware of our motivations and the likely consequences of our actions. To do this, we need to be able to imagine the experience of other people. This allows us to assess the potential impact of our actions. If we have no means to do this, then we have no means to discriminate between what is wholesome and what is unwholesome, between what is harmful and what is not.

So Buddhist ethics is not simply a matter of good or bad. It's the state of mind, or volition, that counts, and the more awareness and imagination you have, the more you are able to act in a skilful way. This comes with practice and experience. It means taking responsibility for the *likely* consequences of our actions, bearing in mind that we don't have certain knowledge about the future. And in the present climate and ecological emergency, it means taking *collective* responsibility for the likely consequences of our collective actions. We could even see it as collective karma, in the sense that humans have performed the same kinds of actions, thereby setting up the same kinds of results.[8]

In chapter 3 we explored a specific basis for Buddhist environmental ethics that emerges from the Pali scriptures. We all depend on suitable living conditions, both now and in the future. Our bodies are dependent on the natural world for our survival. Without suitable living conditions we won't be able to practise and gain Enlightenment. Protecting the natural world supports the possibility of Enlightenment.

This is a helpful start; however, it doesn't seem sufficient for our purposes. Whilst Enlightenment is the most important value in Buddhism, my intuition tells me that the earth and all its creatures are in some way valuable *in themselves*. This brings up deeper questions about the nature of Enlightenment and our relation to the world around us.

Returning to the prose poem at the start of this chapter, perhaps we can find this deeper basis in a sense of solidarity with all life. The universe is alive! In the West this animistic worldview has been obscured by the rationalism of the European Enlightenment and the Protestant Reformation. Remnants survive in raves, ecstatic dance, yoga, Romantic poetry, and paganism. Sangharakshita went so far as to say that 'a universe conceived of as dead cannot be a universe in which one stands any chance of attaining Enlightenment'.[9]

How can this be? We are not merely receivers of messages from a fixed, static universe. The four elements of earth, water, air, and fire are alive, and we are made up of them. We are active participants in consciousness. This calls into question the materialistic distinction between mind and matter, between a living 'me' and a non-living 'not-me'. Alienated from nature, we have lost our affinity with it. We need to recapture the sense that to be human is to be part of nature.

Siddhartha, the Buddha-to-be, grew up in a palace, protected from suffering. His father deliberately sanitized Siddhartha's environment so that he wouldn't be confronted by unpleasantness. Nevertheless, he gradually became aware of the suffering of old age, sickness, and death. In one version of the story[10] he set out on a magnificent horse to see the forest and came across a field that was being ploughed. The surface of the soil was broken with furrows like waves of water.

Stopping for a moment, he noticed that the young grass was torn up and scattered by the plough, and the soil was littered with dead worms, insects, and other creatures. Siddhartha felt grief welling up inside him, as if it was his own relatives that had been killed. Then he saw the ploughman's skin, blackened by wind, dust, and the sun's rays, and the oxen in distress with the effort of pulling. A deep sense of compassion arose in him. Getting down from his horse, he walked slowly over the ground, overcome with grief. Considering how things came into being and passed away, he cried out, 'How wretched!' (We'll return to the topic of climate and ecological grief in chapter 10.)

How could he find peace of mind? His response was to sit down under a rose apple tree and reflect on the suffering he had seen. Reflecting in this way, his mind quite naturally became concentrated and pliable. He experienced higher (dhyanic) states of consciousness. While in this state, a wandering truth-seeker dressed in rags came past and gave Siddhartha a clue for how to find an answer.

Later, when he had found the answer for himself, the Buddha invited his followers to reflect that all beings are terrified of death; to all, life is dear. When we see this, killing becomes unthinkable. Ethics is really to do with feeling solidarity with all life. It's a direct recognition of the life in us, in others. This is clearly a kind of imaginative act, a kind of imaginative empathy. It is direct and immediate, and may be completely intuitive, without thought, instinctive. We simply *resonate* with the life in another person or animal. This is the ethics of empathy, the natural resonance of life for life. In its highest expression it can lead to a spontaneous outpouring of compassion, acting for the greatest benefit of all.

If we pay close attention, we find that we are already sensitive to the life around us. We are most likely to pick up this vibration deep in a forest or jungle. It's not limited to animals or other living organisms. We can be empathically aware of the living quality even in stones or metals, storms or stars. We can sense life vibrating in them.

It can be difficult to feel this empathic awareness living in a city, where the natural world is held at bay. The artificial environment tends to alienate us further and further from the natural world. If we find ourselves in this kind of situation, we can go to parks, notice the seasons, and keep plants. When we have more time, we can go walking or hiking. Anything to get us back in touch with the natural world!

So we can summarize the Buddhist ethical bases of environmental awareness and action as rational self-interest (we don't want to foul our own nest, we want to preserve the possibility of Enlightenment); a sense of solidarity or even identity with the natural world; and finally a spontaneous outpouring of compassionate action for the greatest benefit of all.[11] It is the third of these, a spontaneous outpouring of compassionate action, that David Loy emphasizes in his book *Ecodharma*. He argues that engagement with the world's problems is not a distraction from our personal spiritual practice but can become an essential part of it.[12]

The more people realize that the natural world is alive, full of life that resonates with our own lives, *and is valuable as life*, the greater our chances of survival as a species and a planet. As humans we bear a greater responsibility than other species, as we have been responsible for creating the current climate and ecological emergency, and other species don't have the means to save or protect the world.

Talking about his poem at the start of the chapter, Urgyen Sangharakshita said that we humans should feel that we have done real harm to this planet and its other inhabitants, and that we are still doing harm, and increasingly so. His response was to offer an apology, in the form of a prose poem. He explained that if one has done something wrong, one should apologize. However, it must be a profound apology, and come from the heart, not because someone else thinks that we've done something wrong. Finally, he suggested that there must be reparation for the harm done, a change of heart, a change of direction.[13]

This is Buddhist environmental ethics in action. If we have done something that isn't in accord with how we want to treat others, we openly acknowledge it. We feel sadness and regret because our actions didn't live up to *our own* values – for the safety and protection of all beings. This kind of acknowledgement is an essential bridge to what is most precious to us and paves the way for more concrete reparation or restorative action.

We can express this kind of ethical regret about polluting the air, about destroying the habitats of animals and indigenous people (along with their way of life), about hunting certain animals to extinction or the brink of extinction, about the chemical poisoning of rivers and streams, about the destructive mining of the earth, about continuing to kill whales. Sangharakshita makes the point that some of these losses are reversible, with effort, for instance pollution of certain habitats. Some, like species extinction, clearly aren't.

In the next chapter we'll explore compassionate action based on wisdom, and learn how to work with the states of horrified anxiety that are very likely to arise in the face of the climate and ecological emergency. In the meantime, to strengthen the basis of loving kindness, I invite you to join me in a guided meditation, based on the famous *Karaniya Metta Sutta*.[14]

Guided reflection: the Buddha's teaching on loving kindness

Go to https://www.windhorsepublications.com/audio-resources/audio-resources-for-tbh/

- Purpose of this activity: to radiate loving kindness in all directions and to beings of all kinds.
- Time: 15 minutes.

5

Compassionate action based on wisdom

The single most important factor that will move us to act to protect the world is compassion.

– The 17th Karmapa, Ogyen Trinley Dorje, spiritual leader of the 900-year-old Karma Kagyu lineage of the Kagyu school of Tibetan Buddhism[1]

The climate and ecological emergency is one of a number of problems that the world faces in the twenty-first century. All of these problems have something in common: they are extremely complex.[2] As such, they will require a high degree of cooperation to address. The responsibility for addressing these problems can't rest solely with governments. Even if governments legislate for what their citizens can and can't do, individual citizens will find a workaround using modern technology.

For this reason, our states of mind are even more pivotal than ever before, giving us the ability to cooperate with each other, and ultimately for our survival as a species. The climate and ecological emergency has arisen because we have failed to recognize our interconnectedness and have attempted to meet our own needs at the expense of others. If we see the emergency in this light, and encourage others to do the same, we will have some impact.

The starting point for Buddhist action is, 'If you want to change the world, change your states of mind.' Activists of all kinds say, 'If you want to change the world, change the world.' However, the world and states of mind impinge on each other. Your states of mind affect the people and the world around you. The world affects your states of mind. It's not one or the other, it's both: transforming self and world.

Protecting the environment that we all rely on for our survival is an immediate way to care for all beings. Much of the devastation is the result of consumerism and the technologies of extraction and production that make it possible. Consumerism is what modern society tells us will make us happy and safe and lead to a good life. It's time to rethink that – is it really true? We all want comfort and happiness, but at what cost to others, the environment, and future generations?

Let me give a somewhat tongue-in-cheek example. Despite following the Buddha's teachings since I was nineteen, I've always had a longing for certain types of sports car. When I see a black Audi Five going past me on the outside lane of the motorway, the world stops. In that moment I could only satisfy my needs for freedom, empowerment, and self-respect by possessing such a car. Nothing else will do. I've lost a sense of choice and perspective about my needs. I've narrowed down on possessing this one object, and in the process I've forgotten that I have other options to meet these needs.

To address this loss of choice and perspective, we need to work on our attitudes. As we saw in chapter 3, the Buddhist tradition encourages us to connect deeply with our environment and see the earth as living, perhaps even as our mother, or a goddess. The earth sustains us all: we can't be too grateful for the air and other nourishment that it provides. The earth is our source of life. Where the earth has been regarded as sacred, people have felt a sense of closeness to it and have cared for the environment. Even if we haven't grown up in this kind of culture, it's still possible to develop this attitude of reverence and awe towards the environment as an adult.

As the Tibetan Buddhist leader the 17th Karmapa writes, the single most important factor that will move us to act to protect the world is compassion. This means compassion towards the physical environment and towards the beings that inhabit it. Compassion moves us *to care* and *to act*. Compassion and its companions, loving kindness, empathic joy, and equanimity, are by their nature skilful volitions. We need these kinds of skilful or wholesome emotions, not just sound ideas about the environment.

Compassion is something that can be developed; it's not something foreign to us. We all have it, or the potential for it. The Buddha expressed the conviction that we are able to nurture our minds to become more wholesome, more in contact with reality, and more free. He offered us practices based on this conviction.

So let's unpack compassion and its companions. Loving kindness (*metta*) is the basis for all the others. *Metta* is unconditional love and friendliness towards all beings. It's unconditional because it doesn't depend on other beings returning it. We could talk about it as love, if we are clear that this kind of love has three aspects: cherishing, protecting, and maturing. In cherishing, we hold the other dear, precious, valued. In protecting, we protect that which we hold dear. In maturing, we support and encourage what is precious to us to grow and flourish. This is *metta*.

Metta becomes *karuna* or compassion when it comes into contact with suffering of any kind. The aspect of care and concern comes to the front, along with a desire to alleviate the suffering. When *metta* comes into contact with joy and celebration, it resonates with them and becomes *mudita*, or empathic joy. When *metta* comes into contact with both suffering and joy, it expands to hold these poles, and becomes *upekkha*, or equanimity.

How do we engage with the suffering of the climate and ecological emergency in a sustainable way? According to the Buddhist tradition, compassion is the only sustainable fuel. Everything else will sap our energy.

When we become aware that others are suffering, our natural empathic response is to wish for that suffering to be removed. We feel the urge to remove it ourselves. If we are genuinely empathic, we will not be able to ignore the distress around us. We will do what we can to help. This leads us to compassionate activity in general, to nonviolent social change, and to environmental activism in particular.

It's important to remember that compassion is not just a feeling; it implies the potential for action. The Sanskrit word *karuna* implies compassionate action based on wisdom. Clearly compassion is a central quality or faculty for environmental concerns. It's a congruent response to environmental destruction. It's the wish for an absence of harm, the vision of beings being protected or relieved from that harm, and the actions that result.

It's important to clarify that compassion is not 'taking on the suffering' of the other person. As I pointed out in *I'll Meet You There,*[3] identifying with another person's suffering is actually a block to empathy and likely to lead to distress. Nor is it simply the wish for others to be free from harm. Rather it is an energetic radiation of care and concern in all directions, so that it becomes boundless. The traditional image is of a vigorous conch-blower making a sound that is both all-pervasive and beautiful. In this radiating energy there is no place for harming, and no room for righteous anger, which is likely to be based on ill will.

Many environmental activists are on the point of despair. We need regeneration and support to continue to act. Something that could support us is to understand the nature of compassion more deeply. When I'm considering the climate and ecological emergency, I go through a cycle of feelings, including horrified anxiety, resignation, anger, denial, and moments of compassion. Does this cycle sound familiar? (I'll return to the topic of anger in chapter 9, where we will try to find the life in it rather than pushing it away.)

So, how do we cultivate compassion? First, it's important to remember that *metta* or loving kindness is the fundamental quality of *karuna* (compassion), empathic joy, and equanimity. So *karuna* is a *metta*-full response to anyone who is suffering. There's a sense of closeness, a recognition that other beings are vulnerable to suffering, just as we are. And there's an intense resolve to alleviate that suffering, which translates *directly* into action.

Karuna tends to be a more demanding quality than *metta*. It can be extremely difficult to face our own suffering, let alone the suffering of others. There is a specific meditation practice for developing *karuna*, which we'll return to at the end of the chapter. Paradoxically, in this practice we don't try to develop compassion directly. We simply bring to mind a person or other being who is suffering, and try to develop loving kindness, friendliness towards them. We don't assume that we know what compassion is. We simply include an awareness of their suffering, bearing in mind that this is not the only aspect of their experience.

There's a common misunderstanding that compassion focuses on the suffering of the person. This is very likely to lead to horrified anxiety, as we'll see below. Instead, we focus on the *being* who is suffering and what we can do for them. This is less likely to lead to overwhelm and more likely to lead to action.

We're all aware of the climate and ecological emergency. What does it take for this to shift from our head to our heart? What does it take for something that we know is harmful to become unbearable for us, so that we act? How do we decide to get involved, I mean really involved? We need to experience something that touches our heart and translates our ideas into motivation.

Buddhists have been facing these issues for more than two thousand years. In the tradition we can find tools to support, protect, strengthen, and guide us when we find ourselves beset by cruelty and indifference, or by sentimental pity, horrified anxiety, gloom, helplessness, or weariness. We can imagine a gardener tending to their plants and trees. What does the seed of compassion need to flourish in you? Does the soil need to be moist or dry? Does it need sun or shade? What would it receive as nourishment?

One such tool is the concept of near and far enemies. This tool can help us identify our states of mind and offers antidotes to help us work with them. The 'far enemies' of compassion are cruelty, indifference to suffering, and even inflicting suffering. These far enemies are relatively easy to spot, as they strongly contrast with compassion. In the context of the climate and ecological crisis, indifference can take the form of denial and shutdown.

In January 2020 I visited Melbourne to lead events at the Melbourne Buddhist Centre and train Extinction Rebellion organizers and trainers in nonviolence and de-escalation. My partner Gesine and I were lucky to escape the smoke from the bush fires – the wind was blowing in a different direction. I was struck by how much denial there was in the Australian media, even at that crucial moment. With indifference and denial, what's missing is a sense of closeness with the beings who are suffering, and an appreciation of how similar we are to them and all beings.

The 'near enemies' of compassion are sentimental pity, horrified anxiety, gloom, helplessness, and weariness. Near enemies are harder to spot than far enemies, as they closely resemble the quality we are trying to develop, in this case compassion. However, they are connected. The Melbourne climate activists told me that if they tried to raise their voices to break through the denial, they could easily get into agitation and horrified anxiety.

In either case, we're not really open to the experience of suffering, to being with it. To counteract this, we can imaginatively develop a stronger sense of what we have in common with beings who are suffering, and indeed with all beings. This is what I give the name 'empathic comparison' in *I'll Meet You There*.[4]

There are countless situations that I find deeply disturbing. However, there are two topics that particularly haunt me: people in India who work on the streets, and whales swallowing plastic. (These are just two examples of mine. To get the most from this section I suggest that you substitute your own.) Indians have sat outside their houses, on the road, and engaged in their craft or profession, since the time of the Buddha. Only in the last few decades has breathing the air become hazardous. Nine out of the ten most polluted cities in the world are now in India.[5] I find it almost incomprehensible to imagine the scale of suffering from respiratory illnesses and other conditions that this implies.

I regularly read reports of whales that have beached or been washed up with large plastic objects in their stomachs. I find this horrifying. I keep on replaying the scene of a whale coming up from below, jaws open, to swallow a school of fish. Only, instead of fish, the whale swallows a plastic object, such as a car bumper, that will never be digested or pass through its stomach.

Why do these two images haunt me? I guess because they relate to weaknesses in my body and chronic illnesses that I've suffered in my throat and stomach. With sentimental pity I don't want to feel the shock of my response. With horrified anxiety I become depressed and useless to the world. Both arise from my fear of the feelings that are stimulated in *me* by their suffering.

Sentimental pity involves shying away from discomfort and covering it up by feeling sorry for the suffering beings. Those poor whales! It's as if I'm above the beings I'm feeling sorry for, looking down on them. I'm reluctant to engage with them and their real needs. Pity denies my profound similarity to other beings, and can feed a kind of saviour complex (I'm going to save them!).

The antidote to sentimental pity is twofold: to acknowledge and engage with my actual feelings about the suffering of these beings, and to get involved with those suffering beings. As I write this, I'm curious to note that I feel a strong resistance to doing either of these things!

Horrified anxiety is another form of counterfeit compassion. When I'm feeling it I identify with the suffering, I lose perspective, I stiffen and panic. It's so overwhelming, there's nothing to be done! We're all doomed! In losing sight of the other beings, I am no longer able to be of any use to them. The antidote to this is to recognize the anxiety in my heart/mind and see how it prevents me from really empathizing and being of any use. I could also recognize that I'm not alone in facing this emergency, that I share this responsibility with others.

While I was in Melbourne I met up with Tejopala, a friend and brother in the Triratna Buddhist Order. He's been involved in climate activism since the early 1990s, and is currently working for the Australian Religious Response to Climate Change (ARRCC), where among other things he has been coordinating nonviolent, interfaith protests against the proposed Adani coal mines.

He told me that it's not just me that reacts in this way. In his experience, more than anything else, climate activists tend to get into horrified, overwhelmed anxiety. He notices that he's become more confident and capable himself through practising loving kindness meditation. However, not feeling stressed out has become a bigger and bigger challenge.

Just after Christmas he found out with his partner that she was pregnant. The bush fires were all over the news as they sat in the doctor's waiting room. About three days later they had to drive out of smoke so thick that they could only see about a mile in any direction. Tejopala found himself going to a hardware store to buy respiratory masks for them both.

He paused, 'That kind of thing can press your buttons. Yet even then I think I was not as angry and stressed out or demoralized as quite a few people were at the time. I was used to working with my mind, noticing fear and anger and cultivating a basic love and solidarity with others. Many people were going through much worse, after all.'

Tejopala has taught meditation to climate activists on many occasions; he says that what they really need is loving kindness meditations like the ones that you'll find at the end of this and the previous chapter. He's concerned that most of the available meditation practices tend to be mindfulness-based. In his view what activists really need is emotional positivity. Only loving kindness, compassion, empathic joy, and equanimity can counteract the pull towards sentimental pity, horrified anxiety, and the others.

Do you recall my glider moment in chapter 1? I was able to contemplate my own death only because I was in a very emotionally positive, almost blissful state. And it only lasted a few seconds, before fear kicked in. It's the same here; we need all the loving kindness and compassion that we can muster to look this reality in the face, even momentarily.

In meditation, there are specific antidotes to apply when we notice that we're getting into horrified anxiety. The direct antidote is to develop empathic joy: reflecting on the positive, skilful qualities of someone or a situation, or reflecting on a person's good fortune and happiness as something to recognize and rejoice in. This can help to restore balance and gain perspective.[6]

So when I'm feeling horrified, overwhelmed anxiety in relation to the immensity of suffering in the world, and in particular Indians working on the streets and whales swallowing plastic, I can take a step back and see all the amazing things that people are doing to relieve suffering. I can remember that individual efforts do have an effect, and that there are wonderful people in the world, like my friend Tejopala, who have been active for a lot longer than me, who are doing amazing things.

When Tejopala realized the true scale of the climate and ecological emergency in 2012, he asked himself, 'What's the most effective use of my time here on this earth?'[7] I can rejoice in the sense of meaning and purpose that answering this question has brought him, and the contribution to the world that has come from this.

Does it feel different in the pit of my stomach? Somewhat – I notice that I feel hopeful alongside the tension and heaviness. Antidotes have that kind of effect; they can lay another feeling alongside the original feeling, without completely transforming it. My preferred approach is self-compassion, which offers the hope of genuine transformation, rather than consolation.

The process for self-compassion which I've learned from Nonviolent Communication[8] involves identifying a specific observation about the topic, sensing the feelings that arise in my body, connecting to the needs that underlie those feelings, and finishing with a specific commitment to act to meet those needs. I'll follow through this process for myself in relation to whales swallowing plastic.

When I recall the number of reports I've read of plastic being found in whales' stomachs (observation), I feel sick, disgusted, and heavy (feelings). If I explore underneath or behind these feelings, I find that I am longing for peace of mind, and to know that all animals on the earth are safe and protected (needs or values). When I connect to these needs/values, I notice that now I'm feeling sad rather than sick and disgusted. I take this as a sign that I am really connected to my needs/values. To complete and ground the process, I imagine a range of next steps I could take to protect animals on earth, and I select three specific steps: to spend time this week learning more about the issue, to include this example in this chapter, and to talk about it when I'm doing book launches and workshops based on this book. If I don't mention it when you see me, could you remind me?

What's the impact of this kind of self-compassion process? I definitely feel relieved, as I'm going to be *doing* something, rather than just letting the anxiety gnaw at my innards. At the same time, I'm aware of a deeper anger and grief. I'll return to these feelings in chapters 9–10 and explore nonviolent social action in chapters 12–14.

How do you care for your own feelings of horrified anxiety? Do you think that self-compassion could work for you? You might like to try it on your own or with a partner/friend. Buddhism has a richness of approaches to working with the heart/mind. You could also try: connecting to an inner resource; allowing love in; cooling it down; unhooking; reflecting on conditionality; developing a broader perspective; and going for refuge or calling on a protective figure. All of these have worked for me at different times to different extents. Some are part of my daily meditation practice.

Connecting to an inner resource. Find a place in your body that feels safe or comfortable. Enjoy this sense of comfort and safety for a moment. Then bring to mind something that would ordinarily stimulate horrified anxiety (in my case, whales swallowing plastic). Then come back to your inner resource again. Go back and forth between the triggering image and the inner resource. If it doesn't work for you to find a safe, comfortable place in your body, you could try bringing an inspiring figure to mind as your inner resource.

Allowing love in. This is one of my favourite approaches, probably because historically I've lost emotional warmth very quickly when I've encountered difficulties. I just say, 'Allowing love in . . .' and bring to mind my partner Gesine, my parents, all beings, my teachers and brothers and sisters in the Order, my Buddhist and NVC students. Then there's the Buddhist figures that I have a particular connection with, Green Tara (see below), Padmasambhava (the mythical bringer of Buddhism to Tibet), and Avalokitesvara (Bodhisattva of compassion). That's Team Shantigarbha! I allow love into my heart from these sources, being careful not to force anything, just allowing it in where I can feel it. And then I allow love through me, so that I become a node or a conduit for love to pass from beings to other beings. This really soothes my heart.

Cooling it down. Horrified anxiety tends to be a hot emotion – you fix on a doom-laden vision of the future and become completely engulfed and entangled in it. To cool it down a bit, you can bring in an element of uncertainty. We all want the comfort of certainty about what will happen in the future; however, there is none. So you can remind yourself that we don't *really* know the impact of the climate and ecological emergency. We can make pretty accurate scientific predictions, but we don't really know how it will impact our lives or the lives of others. Try this out very subtly at first, and stop if you find that it just increases the heat!

Unhooking. This is a sheer cliff face of a practice for moments of extreme tension. Pay attention to the bodily sensations associated with the tension. Focus on one of these sensations. Stories will come into your mind about this sensation. They will present themselves as cogent reasons for this sensation, or what this sensation is about. Each time a thought or a story arises, unhook the story from the sensation. You can imagine yourself physically unhooking something that's trying to attach itself to the sensation. Keep focusing on the sensation. Keep your attention there, until the story quietens down enough to use other approaches. Phew!

Reflecting on conditionality. Earlier I suggested that denial, cruelty, and indifference are the 'far enemies' of compassion, whereas sentimental pity, horrified anxiety, and the others are the 'near enemies'. To understand the differences between them more deeply, it might be helpful to relate them to the Buddha's teaching of the Middle Way.

There were two metaphysical views that were very popular among religious seekers in the time of the Buddha: eternalism and nihilism. Eternalism is the tendency to believe in heaven, some kind of god, or an eternal soul that passes from body to body. Nihilism is its opposite: the belief that 'There is nothing!', a rejection of any form of eternal truths, and a rejection that actions have consequences. The Buddha pointed out the limitations of both these views, putting forward a Middle Way characterized by dependent arising – everything arises in dependence upon conditions.

Denial and indifference tend to be associated with an eternalistic mindset. We can see this in the argument put forward by some Christians that God's will is being fulfilled on earth, so there's no emergency and nothing to be done about it. Horrified anxiety can result in nihilism, which focuses on endings and tends to lose sight of meaning and value. Integrating and transcending these two views is compassion, which is in accordance with the understanding of dependent arising. All phenomena arise in dependence on conditions. When those conditions cease, those phenomena cease. In fact compassion is the reflex of this kind of understanding. When we see how deeply interconnected we are with all beings, compassionate action spontaneously arises to relieve suffering.

Developing a broader perspective. What's the bigger picture here? In your mind's eye, pan out to samsara or conditioned existence. This is something we're going to explore further in chapter 7. We know that suffering is an aspect of samsara. Eco-suffering is a part of this. Can we embrace an awareness of eco-suffering? There are parallels here with embracing the reality of death – it can lead to a renewed sense of the preciousness of life. From this broader perspective, commit yourself to a specific action to alleviate the suffering.

Going for refuge/calling on a protective figure. If none of the other approaches seems to work, and you have faith in the Buddha, Dharma, and Sangha, you can recite the refuges and precepts and commit yourself to following them with body, speech, and mind. If you have a connection to a specific Buddha or Bodhisattva, you can call on them by reciting their mantra. I have a connection to the figure of Green Tara, a beautiful Bodhisattva associated with active, engaged compassion. She is shown with one leg folded in meditation, resourcing herself. The other leg is stepping down, ready to take appropriate action in the world. She is known to be a protector. In Tibet travellers and others call on her to protect them from dangers of all kinds. Her compassion is always accessible; there are no special conditions to fulfil.

Humans don't own the earth; no one does. However, we can be guardians, holding the planet in trust for future generations. We have intelligence, we have an ethical sense, and we can nurture compassion. We need to bring these three faculties (intelligence, ethics, and our emotional response) together for the benefit of all beings, especially other animals and our environment. Our relationship to the environment will become truly sustainable when this happens.

In the next chapter we will explore the power of intention by looking at the five precepts (training principles) of Buddhism in relation to the climate and ecological emergency. With compassion in your mind and heart, what specific next steps are you willing to commit to? But first, I invite you to join me in a guided reflection to nurture compassion towards the environment and the beings who rely on it.

Guided reflection: nurturing compassion towards the environment

Go to https://www.windhorsepublications.com/audio-resources/audio-resources-for-tbh/

- Purpose of this activity: to nurture compassion towards beings who are suffering as a result of the climate and ecological emergency.
- Time: 20 minutes.

6

The five precepts

> Behold, I do not give lectures or a little charity,
> When I give I give myself.
>
> – from 'Song of Myself' by Walt Whitman[1]

Now that we have explored the importance of compassion in moving us to act to protect the world, we're ready to harness the power of intention. Intention contains the seed of what you want to grow. We are more likely to follow through with things that we are very committed to. So dig deep. Ask yourself: what is your deepest heart wish in relation to the climate and ecological emergency?

Is it that all beings are safe, both now and in the future? Do you want current and future generations to have clean air to breathe, clean water to drink, and space to enjoy the bounty of nature? Whatever your heart wish, I invite you to make a conscious intention to act, to go in a particular direction.

As we saw in chapters 4 and 5, motivation really matters. Dos and don'ts can easily become a set of rules. As you try to follow these rules, it's easy to become a kind of eco-bore.

However, any kind of morally superior attitude (as we'll see in the next chapter) will fail to inspire others to take action. Rules invite either submission or rebellion. You either submit and lose your autonomy, or rebel against them, keeping your autonomy but paying a price in terms of belonging, support, and so on. So I invite you to choose a first step on the path as a magnet that can draw you on, rather than in a goal-oriented way.

So where to start? The Buddha outlined five basic ethical principles or precepts. They are recited as part of a ritual known as going for refuge, in which one affirms that one is a Buddhist, though you don't have to be a Buddhist to value or practise them. They are guidelines to help us cultivate nonviolent and loving states of mind, to become more Buddha-like. I've come to see them as training principles, each with an emphasis on self-discipline and awareness, rather than moral commandments. They apply to body, speech, and mind, as whatever we do in these three areas will have a karmic effect. And each principle can be broken into smaller steps, also called precepts.

The precepts embody a desire that other beings not come to harm. They can be expressed either in terms of what to avoid or in terms of what to practise, cultivate, and develop.[2] We can use them to explore what a first step might look like. What specific precept can you take in this area? What can you commit to?

I undertake to abstain from taking life.
With deeds of loving kindness, I purify my body.

The first precept involves abstaining from taking life and injury to life. Traditionally this meant casting weapons aside and being conscientious about not depriving living beings of life. Killing is nearly always doing something to them that they don't want you to do to them. It's nearly always the last thing that they want you to do to them. It can only be achieved by using 'power over', in the most basic situation by using your greater physical strength to take away life.

Killing is a total denial of being. Practising this precept means abstaining from killing and more, acting in ways that affirm being rather than denying it. It means giving space to other beings to allow them to be themselves. It means not taking advantage of others because you have more knowledge or experience. It means not trying to force or manoeuvre a situation. It means generating loving kindness and acting from this state, whether you like or agree with others or not.

Through accepting this precept we recognize our relationship to all life and realize that harming any living creature harms oneself. By devaluing the life of another, we devalue our own life, our own sense of humanity. As I mentioned in chapter 1, life is dear to all beings. When we compare our own experience with their experience, it becomes unthinkable to kill or cause to kill.

This first precept introduces us to the principle of *ahimsa*, which is the foundation for all five precepts. The Sanskrit word *ahimsa* is made up of the prefix 'a' meaning 'not', and 'himsa' meaning violence. So we can translate *ahimsa* as nonviolence or non-harm. However, nonviolence is not just the absence of violence and killing. In its fullness, it is the presence of care, kindness, love, and compassion – the urge to reduce suffering.

So the first precept can be applied to areas of life including food, land use, pesticides, pollution, and even how cultural and economic systems spread from one part of the world to another. When natural habitats are removed, it has an impact on the native animals, on local people, and their current ways of life. When we pollute the water, it has an impact on the flora and fauna that are dependent on it for their survival. Bear in mind that from a karmic perspective non-action is also a kind of action. If we are aware of opportunities to stop the killing of living beings and don't take them, this will also have an effect on us.

How do we live in the environment without harming others or endangering their lives? How do we cultivate reverence for life while living in a consumer culture? The Buddha gave his forest-dwelling followers guidance that was remarkable in its environmental consciousness. What could be our modern, post-industrial equivalents?

Speaking from the heart, the most effective way to protect life is to move towards a plant-based diet. What we eat has a direct impact on our ecological footprint. Methane from livestock has a greater impact than worldwide CO_2 emissions. Plant-based foods use land and other resources more efficiently. All this besides reducing the suffering of meat production and slaughter. To make this kind of change, we need to connect with something that moves our heart. Could the environmental impact, on top of the basic reduction in animal suffering, be enough?

Buddhist ethics isn't all or nothing; it's about the next step you can take to act just a bit more skilfully. What's your next step, what can you commit to? Perhaps it's eating a plant-based diet for one meal a day, or one day a week. Perhaps it's committing to a plant-based diet for a period of time, say a week or a month, or three, or forever. As in all ethical matters, this is not an excuse to judge others. For example, some people won't be able to take these steps towards a plant-based diet for health reasons.

One thing we can all do is to cultivate loving kindness directly, in meditation. As we saw in chapters 3 and 4, we can radiate loving kindness towards all creatures in all quarters, without restriction. We can reflect that life is precious to oneself and others, and cultivate a reverential attitude towards all forms of life.

If we notice that we're feeling resentment or ill will towards politicians we assess as 'doing nothing', fossil fuel corporations, loggers, and others whom we regard as responsible for increasing the global threat, we can radiate loving kindness in their direction. This doesn't mean that we agree with their worldview or their actions. It means that loving kindness in its fullness is unconditional, for all beings.

I undertake to abstain from taking the not-given.
With open-handed generosity, I purify my body.

The second precept talks about abstaining from what is 'not-given'. This is more than not stealing; it involves careless borrowing, embezzlement, fraudulent business dealings, and so on. More generally it includes the discounting of other beings and a lack of respect for their dignity and needs, including their property.

Applying this precept to the climate and ecological emergency takes us to the topics of global trade and the corporate exploitation of resources. The Global Footprint Network estimates that we've been using the earth's renewable resources in an unsustainable way since the 1970s. By 2008 this overuse had increased to the equivalent of the resources of 1.5 earths.[3] This means that it takes 1.5 years for the earth to regenerate the renewable resources that people use, and absorb the CO_2 waste they produce, in that same year. Clearly this is unsustainable; those resources will eventually be depleted.

This gives us a lot to reflect on. In relation to the environment, how and when have I taken the not-given? How do you determine what is 'given'? What are my needs, as distinct from my wants? According to the Chilean economist Manfred Max-Neef and his colleagues at the economic school of Human Scale Development, there is a list of fundamental human needs.[4] These needs are finite and classifiable. Wants on the other hand are infinite.

Max-Neef and his colleagues regard these fundamental human needs as being common to all humans throughout time and culture. They are: subsistence, protection, affection, understanding, participation, leisure, creation, identity, and freedom. Max-Neef added the possibility that a tenth, transcendence, may in time become a universal need. Humans, and indeed all beings, are always trying to meet their needs. We are always trying to enrich our lives, sometimes wisely, sometimes unwisely, sometimes taking into account the needs of those around us, sometimes not.

What are humanity's needs in relation to the earth's oil and coal? I'd like to suggest that we humans are taking the not-given on a planetary scale. Who are we taking these resources from? Those human and non-human beings who need them, now and in the future.

We need to review our participation in unwise marketing and consumption, in both goods and services, especially when future options and heritages are being irreversibly removed from future generations. This amounts to stealing from the unborn. Are we doing our part to ensure the wise development, marketing, and consumption of goods and services?

Inevitably the impacts of such a global overuse will be more serious for those with fewer resources, less of a cushion against adversity. So care for all beings entails care for their environments and for climate justice. A Buddhist scripture argues that a primary cause of theft and violence is poverty. Punishment will not change society. The Buddha figure in the story says that to stop crime you need to improve economic conditions.[5]

With these reflections in mind, what's your next step, what can you commit to? We could all consider taking precepts in any or all of these areas: reusing, recycling, how we shop, how we consume, how we travel, how we clean ourselves, how we heat our houses, how we invest, how we reconnect with the earth, how we garden, how and where we live, and so on.

Take your pick! It would be very easy to feel overwhelmed at this point. This is a pitfall to look out for. One way to avoid this is to identify one or two areas that you feel most motivated to commit to, research them, and form a personal precept that reflects what you can realistically achieve and what will have some impact. There is a skill to pitching a precept so that it's not too hard, and at the same time it's not too easy, so that it stretches you.

For instance, when our car was taken from the street without our consent last year, we made the resolve that our next car would be at least partly electric. Now we have a plug-in hybrid. We're finding that it can be difficult to charge it when we're away from home; we're learning how to use the different charging systems in the UK. At the same time we're happy to be relatively 'early adopters' to support the growth of this technology and infrastructure.

I undertake to abstain from sexual misconduct.
With stillness, simplicity, and contentment, I purify my body.

The third precept starts with refraining from exploiting or hurting others sexually. Sadly, most sexual violence is directed towards women, so we need to look at societal attitudes. How do attitudes towards women encourage or enable sexual violence towards them? In recent years I have been involved in many conversations about unwanted sexual advances towards women in public spaces in India. I invoke the wisdom of Dr Ambedkar, Bodhisattva hero of the Dalits, who said, 'I measure the progress of a community by the degree of progress which women have achieved.'[6]

However, the positive precept points us in a larger direction. In contemporary life stillness, simplicity, and contentment aren't highly valued. From a Dharmic perspective, they are profound manifestations of the inner attitude of Enlightenment.

Of all your life decisions the one that will have the biggest impact on your carbon footprint is the decision to have one less child. A study published in 2017 found that having one less child can save an average of 58.6 tonnes of carbon for each year of a parent's life. The figure was calculated by totting up the emissions of the child and all their descendants, then dividing this total by the parent's lifespan. Each parent was ascribed 50 per cent of the child's emissions, 25 per cent of their grandchildren's emissions and so on. The next biggest carbon saving you could make is by going car-free, saving 2.4 tonnes.[7]

Without ennobling enforced poverty, it's worth remembering that the Buddha and his wandering disciples lived lives of extreme simplicity compared to most of us in the West. Once a visitor claimed that a certain great king was happier than the Buddha. The Buddha replied that he could sit still in meditation with happiness for an entire week. By contrast, the great king couldn't happily sit still even for a day.[8]

We need to take into account the profound impact of consumerism on our attitudes, beliefs, and lifestyles. Consumerism is the belief that we can rely *totally* on things, material wealth, to provide happiness and avoid pain. Now of course it's not material things in themselves that are the problem here, it's our unrealistic expectations that they will give us the happiness that we seek. Given the all-pervasive nature of consumerism, it's a struggle for all of us, even followers of the Buddha, to free ourselves from this limiting story, to live a life of depth and meaning.

Modern researchers have confirmed the limited extent to which consumerism can produce happiness. In North America happiness surveys have been going since the Second World War. I was fascinated to read that happiness ratings increased then peaked in 1957. Since then, people have got no happier. If anything, they are slightly less happy. However, income and consumption levels have more than doubled in the years since 1957. In the years 1986–1994 people's estimates of what income they would need to be happy more than doubled.[9]

How could this be? There seems to be a link to working more. In the UK and USA working hours have risen, not fallen. Then there's research that says that we quickly adapt to material gain. It also makes a difference *who* we compare ourselves with. Our happiness is largely based on the comparison we make with the people around us. Parents with children will know this. Children tend to compare themselves with their peers. If their peers watch TV, or are playing with the latest gadget, then this is how they will compare their happiness.

For us adults it's the same, only more diffuse. Take for example the people of East Germany. Although they became wealthier after unification, their self-rating of happiness fell. It turned out that they were comparing themselves with different groups. Before unification they contrasted their lot with other Soviet bloc countries and found that they were doing relatively well. So they felt happy. After unification they began measuring themselves against the West Germans, on the TV, in advertising, and in real life. They found that they weren't doing as well as the West Germans, so they felt less happy, despite their increased income.[10]

Happiness research confirms what we know from the Dharma, that happiness doesn't come from material possessions. It comes from meaningful engagement with life, from the positive values that we live by, and from the quality of our relationships, especially our care for others. You might like to try this brief reflection. Think of two or three moments in your life when you were most happy. What gave rise to that happiness? To what degree did that happiness involve buying or consuming things?

So what can you commit to in the area of the third precept? How could you experiment with simplicity and reduce input? Less screen time? Switching your phone off for periods each day? (My phone is switched off as I write this – I simply can't concentrate with it on.) Periods each day when you are away from devices or a full 'digital detox'? Without dismissing the joys and learning that come from travel, could you learn to love the place that you're in, rather than moving around when you feel restless? How could you simplify your food?

I undertake to abstain from false speech.
With truthful communication, I purify my speech.

The fourth precept is not just about telling lies – deliberate falsehoods. It includes avoiding harsh speech, backbiting, and idle gossip. The positive formulation gives us a direction for what to practise: truthful communication whether written or spoken; not to deceive or manipulate ourselves or others. But do we really know what is true for us? I find myself and my motivations opaque, and I've heard from others that this is also true for them. Truthful communication involves a thorough self-examination, self-awareness, and, ultimately, self-knowledge.

Areas where this precept can be applied in society include advertising, consumerism, corporate PR, government propaganda, and fake news. I invite you to reflect on how you talk about the environment; do you communicate reality? In your community who speaks clearly and honestly about the climate and ecological emergency? Do the media? If they don't, do you challenge this? Do schools speak clearly and honestly to their students? If not, do you speak up for a more truthful education?

As early as 1988 Urgyen Sangharakshita was encouraging the Triratna Buddhist Order and Community to take a much stronger stand on issues of this sort, and to play a more active part, at least in our individual capacities, in the environmental movement. In this way, the Order would develop an ecological dimension.[11]

Twenty years later he encouraged the Order to raise its voice against things that are harmful for the world and for the perpetrators of the harm, that is to say, us human beings. He pointed out that acting unskilfully is also harmful for the authors of the acts.[12]

In support of this, in October 2019 the council and community of Taraloka Buddhist Retreat Centre for Women declared an ecological emergency. Taraloka, founded in 1985, is situated in a beautiful part of the Welsh Borders countryside. They cited the 2018 IPCC report, which stated that without dramatic action we will not be able to limit global warming to a 1.5°C increase,[13] and the 2019 UN biodiversity report, which estimated that around one million species already face extinction, many within decades.[14] The statement says, 'Planet Earth is at present under severe ecological threat.'[15]

I asked Maitridevi, the chair at Taraloka, what motivated her. She told me that it was very particular and personal. 'It was down to our swallows. I'm a very keen naturalist, so I really notice what's around me. I moved to Taraloka five years ago, and this is the first place where I've lived where swallows come every summer and nest. Usually there are eight or nine nests, and I look at them regularly, and get woken up at half-three in the morning by them singing outside.

'Three years ago they had a bad summer, with poor weather conditions. The next year five pairs turned up, then last year two turned up but didn't nest. This year, zero nests again. So we completely lost our swallow population within the space of two years. Personally I felt completely devastated by that. Although climate change is out there, it can still seem abstract, but this was totally concrete. I didn't think that I would live in a world without swallows. Suddenly that became very real. Swallows have been in chronic decline since the seventies, as have many species. So when they have difficult weather conditions, they don't bounce back. I realized that they may never come back. That really upset me. We are changing the world so drastically that there is this kind of brittleness or fragility in so many species.

'I find this biodiversity issue really scary, so I wanted to say something. I was involved in conservation for years before I became a Buddhist. Buddhism and conservation are so obviously connected. I wanted to say that this really matters.'

The Taraloka council and community reflected that as Buddhists they have a responsibility to meet the situation with love, awareness, creativity, and wisdom, and to alleviate suffering where they can. Recognizing that it is an unprecedented situation and that no one has the answer, they sought to make their response 'adequate to the situation and to our practice of the Dharma'. However, declaring an emergency for them wasn't synonymous with panic and overwhelm, or a collapse into nihilism. Rather, they were saying that this issue requires urgent attention, clarity, and decisive action.

There are many other climate and ecological statements by Buddhist groups, schools, and movements.[16]

What can you do in relation to the fourth precept? As well as speaking truthfully about the environment, you can talk to the people in power, writing to your elected representative, to local and national authorities, using social media. You could undertake to educate yourself about a specific issue, as I've undertaken with plastic in whales' stomachs and respiratory conditions in India. You can join an existing campaign or even start your own!

I undertake to abstain from taking intoxicants.
With mindfulness clear and radiant, I purify my mind.

The fifth precept has traditionally been interpreted as refraining from consuming intoxicating substances such as alcohol and drugs, including involvement in production and dealing in these substances. It recognizes the impact of drink and drug use in the short and long term. However, when we come to apply this precept today, it's not just drink and drugs that can cloud the mind. It's all forms of addiction: TV, computer games, junk magazines, and screens of all kinds.

If you want to reflect on this precept, you could ask yourself: do you over-consume things that are harmful to the environment? Are you addicted to consumption that is harmful? Do you turn a blind eye towards others who are addicted?

Are you aware, for instance, of the tropical deforestation and water and soil pollution from the cultivation of narcotics such as cannabis, coca, and opium poppies? The environmental damage is severe because crops are often planted in fragile forest environments and follow slash-and-burn clearance. The chemicals that are used at all stages of production have a harmful impact on tropical ecosystems and human populations.

In its positive formulation, the precept rejoices in mindfulness, or awareness. Awareness is crucial. Along with friendliness, awareness lies at the heart of Buddhist practice. With awareness of how we're currently choosing to act, and its likely consequences, comes choice about how we might act differently in the future. In relation to the climate and ecological emergency, awareness means awareness of the likely consequences of our *collective* actions.

Awareness really needs a whole chapter to itself, so we'll be returning to it in the next chapter. In the meantime it might be worth cultivating awareness of another aspect of our current predicament. You might have noticed, as you've been delving into the climate and ecological emergency, a kind of disintegration of identity. A loss of a sense of who you are. As we go into the burning house you might find yourself reassessing your priorities. What is most precious to you at this time? It could be a moment to sit in not-knowing.

Guided reflection: sitting with your shit
Go to https://www.windhorsepublications.com/audio-resources/audio-resources-for-tbh/

- Purpose of this activity: to sit still and notice whatever is going on inside you, without physically moving.
- Time: 15 minutes.

7

The burning house

Let me now make my meaning clearer with a parable.
Through parables intelligent people can reach understanding.

– The Buddha in the *White Lotus Sutra*[1]

Welcome to the burning house. Did you feel the heat as you came in? You might have come here because you heard Greta Thunberg say, 'I want you to act as if the house was on fire.'[2] Or maybe you love the Buddhist parable of the burning house and you want to find out how it applies to the climate and ecological emergency. In any case, welcome!

In this chapter we're going to explore the role of mindfulness or awareness in relation to the climate and ecological emergency. In this context awareness means awareness of the current situation and the likely consequences of our collective actions. In chapter 5 we looked at the opposites of awareness: denial or simply ignoring the threat. The parable of the burning house from the *White Lotus Sutra* is about people who are ignoring an overwhelming threat. Symbolic stories can help us to think creatively about the situation. As the

Buddha says, through a parable intelligent people can reach understanding.

So let me tell you the story of the burning house. Once upon a time there was a very wealthy man who had many children. In fact, he may have been father to up to thirty children. They lived in a huge wooden mansion with only one entrance. But the mansion was old and decaying and full of ghosts and flesh-eating demons of many kinds.

One day the father went out for a while. When he returned, he saw that the house was on fire. On all sides he could see flames consuming the ancient posts, roof beams, and timbers. The demons took to attacking and killing each other and everything was in confusion and distress. To make matters worse, a bystander told the father that all his children were still playing inside the house.

The story takes us inside the mind of the worried father. He is still strong and able, and he wonders about carrying his children out of the house, one by one. But looking again at the burning timbers, he realizes that it will be impossible to get them all out in time. So he cries out to them, 'The house is on fire! You're in terrible danger! Come out quickly!' But the children don't recognize the danger that they're in, and carry on playing.

By now the father is distraught. It's clear that this approach isn't going to work, so he considers his options. He has a brilliant idea. He knows what kind of toys each of his children loves. What if he tells them that their favourite toys are waiting outside? So he calls out, 'There are some wonderful toys outside! You can't imagine how wonderful! They're just outside. You can play with them as long as you like!' The children instantly drop their old toys and games and rush outside, jostling and scrambling to get out first.

Once the children are safely outside, they start clamouring for their new toys. It turns out that the father's wealth is more or less inexhaustible. Instead of the variety of different toys that he has promised, the father presents them all with the same toy, but one that is more wonderful than anything they could ever have imagined. The children are overjoyed and play with their new toy to their heart's content.

So that's the parable. What do you make of it? In its original context the Buddha tells the parable to explain how three earlier teachings (the old toys in the story) can be superseded by a later and more universal teaching, amidst the fire of greed, hatred, and delusion. Clearly this was a problem for the author(s) of the *White Lotus Sutra*.

The concerned and skilful father is of course the Buddha, and his mansion represents the whole universe. Despite its size the mansion is extremely rundown, ready to collapse at any time. This reminds us that the world is impermanent and insubstantial, and therefore not likely to give us satisfaction in a reliable manner. Besides being rundown, the mansion is haunted by ghosts. Is our world haunted too? We're haunted by our past. In particular, we're haunted by our individual and collective actions that have led to this climate and ecological emergency and social and economic injustices too. The ghosts in the mansion are the ghosts of the past that we carry in our minds all the time. We could say that our past has caught up with us.

In the parable the mansion catches fire at a certain time. In reality the world is constantly burning and blazing with the fires of greed, hatred, and ignorance. The Buddha goes into this on another occasion, when he talks to hundreds of matted-haired ascetics. Until recently they were fire-worshippers who spent their days performing fire rituals. Now the Buddha tells them, 'The whole world is ablaze. The whole world is burning with the fire of craving and desire. It's burning with the fire of hatred and aggression. It's burning with the fire of ignorance, delusion, bewilderment, and lack of awareness.'[3] In other words, we don't need to go looking for fire, or to treat it as sacred. The fire is all around us. Nowadays it's horribly visible in forest fires, combusted carbon, nuclear bombs, melting ice caps, and rampant consumerism.

The fire is a threat to the children in the mansion. They are in danger of being burned to death. The children of the parable represent us. We are in danger of suffering endlessly from the consequences of greed, hatred, and delusion. And literally all beings on planet earth are in danger of suffering from rising temperatures and their knock-on effects.

The father considers carrying his children out one by one, but soon realizes that he doesn't have time. Besides, from the point of view of the parable, you can't force beings to evolve against their will. People evolve because they want to evolve, and are motivated more by positive inspiration than by force. So the father tries calling them instead. This represents the call of the Truth.

Many of us have heard such a call at some point in our lives. It's the sound of silence, the call of the Beyond, the sense that there must be something more to life than this. Usually we ignore this call, preferring our familiar comforts. Or we're too busy enjoying ourselves even to hear it. Like the children, we are absorbed in our games: buying stuff, being successful, making sure that we're loved, being right. We run from one game to another in a rather restless and anxious way. As we run, we throw the occasional distracted glance in the direction of the Buddha.

The father knows which children love which toys, so he promises them their favourite toy. This kind of empathic offer is known in Buddhism as *upaya kausalya* or 'skilful means'. In the traditional parable the father promises his children different kinds of carts. These 'vehicles' correspond to different forms of the Buddha's teaching (also known as vehicles), or even different forms of Buddhism adapted to different temperaments. This suggests that a particular appeal will be more effective than a more general one.

The sutra asks if the Buddha's skilful means amounts to a white lie, a form of false speech. The sutra responds that it isn't lying: the father saves his children's lives, and in the process they all get far more than they were expecting. In more modern terms we could describe it as the protective use of force, which I'll return to in chapter 14.

The Buddha uses skilful means and empathy to get his children out of the house. Once the children are safely outside, their father gives them all the very best kind of vehicle, bigger and better than anything they could have imagined. It's as if he promised them a pair of skates, a skateboard, or a scooter, and what they find waiting outside is their first (all-electric, 100 per cent recycled!) sports car. Mine's a black Audi Five![4]

As activists and people concerned with the climate and ecological emergency, we might be tempted to see ourselves as the father, calling in vain to his children. After all, we can see a danger that others can't. However, it's much more likely that we are the children, and scientists are taking the role of the father calling in vain.

The father in the parable gives a cry of anguish and then a cry of inspiration. Scientists are able to give a cry of anguish. They are largely unable to give a cry of inspiration. Why is this? Since the European Enlightenment, physics, the study of the natural world, has been separated from metaphysics, the branch of philosophy that deals with the first principles of things, including abstract concepts such as being, knowing, identity, time, and space.

Science has made great progress by limiting itself to observables and reproducible results. However, this has happened at the cost of being divorced from a realm of broader meaning. As Albert Einstein said, 'It would be possible to describe everything scientifically, but it would make no sense. It would be a description without meaning – as if you described a Beethoven symphony as a variation of wave pressure.'[5] So we still need a cry of inspiration.

Let's take a step back for a moment. What happens when the fire alarm goes off? People tend to sit still and look around nervously to see if anyone else is taking it seriously. It's time to break the spell. If we take it seriously and run out of the house, other people will do the same.

Why is it so difficult for us to hear the alarm, especially when we're aware that we're heading for destruction? One reason could be that we are creatures of habit and comfort. I know from my own experience that I get into a certain activity, a certain habit, a certain way of life, and if I'm not careful at some point I lose awareness. I lose a sense of choice about what

I'm doing. It's easier to continue than stop. Stopping would mean having to acknowledge discomfort and dissatisfaction, including how I feel about the likely consequences of my actions. It's easier just to keep going.

If we want to address the climate and ecological emergency, we're going to need a collective change of heart. This requires a social fabric and community that helps people experience a change of heart. It begins when we recognize each other's needs as valuable. As in Covid-19 times, it may include re-evaluating such activities as care and teaching. More broadly we can recognize and accept the validity of the needs of other species, and acknowledge the claims of future generations for clean air, clean water, and a habitable earth.

As we know from the parable, it's not enough to tell people that the house is on fire and appeal to their sense of self-preservation. And we can assume that they have already heard an appeal to 'good behaviour' (altruism) and are ignoring it. As we saw in chapter 4, a morally superior attitude will fail to inspire others to take action. So we still need a cry of inspiration.

The father promises wonderful toys, exactly what each of his children would love. What could be our modern equivalent? It needs to be a cry of inspiration, not just risk aversion. Could it be a vision of interdependence, a felt awareness of living in a shared world? A call to nonviolence towards living beings, just because they are alive and sensitive like ourselves? A reminder of human potential, a call to a collective change of heart, and a community that helps us to experience this? Are we back to wisdom and compassion, the virtues of Enlightenment?

Let's come back to the starting point of this chapter, exploring awareness in the context of the climate and ecological emergency. What can we become aware of? For instance, when did you first notice the smoke? If the house was really on fire, we'd be looking for firefighters, emergency strip lighting and fire exit signs. In other words we'd be looking for help and guidance on how to get the hell out of here. At this point we wouldn't be looking for fire safety consultants to reduce the risks. It's not even that we can smell the smoke, it's that the flames are now rising around us. Extreme weather events, such as the fires in Australia and California, have become more common. In Europe increased temperatures are having an impact on agriculture. Warming has been locked into the oceans. As the Arctic loses its ice covering, this contributes to further warming.

I first noticed the smoke in India: a poisonous concoction of kerosene cooking fumes, vehicle exhaust, and industrial emissions. I could finally scale up the issue when I arrived for the first time in Bangkok in January 2019, the week that it became, temporarily, the number two most polluted city in the world. Walking the streets between *wats* (temples) and catching water buses along the *khlongs* (canals) I felt a burning sensation in my throat, tightening in my chest, and dizzying nausea. Scaling this up, I finally understood how human activity on a global scale could contribute to the rise in greenhouse gases.

How about you? When did you first notice the smoke?

Secondly, when did the fire alarm *actually* go off? Clearly it didn't start in 2019 when Greta Thunberg told the EU Parliament that she wanted them to act as if the house was on fire. In chapter 2 we saw that scientists have been trying to quantify the human influence on climate for decades. There was the 1999 paper asserting that the changes that they were seeing in the surface temperatures were *largely* anthropogenic. In fact natural factors had contributed very little, particularly since 1950.[6]

The statistician for the 1999 paper, Professor Myles Allen, contributed to the third IPCC report in 2001, which concluded, 'There is new and stronger evidence that most of the warming observed over the last 50 years is attributable to human activities.'[7] This didn't have much impact at the time, and 9/11 happened soon after, diverting political attention elsewhere.

The IPCC's 2007 report got a very different reception – Al Gore's film *An Inconvenient Truth* came out at the same time. However, the conclusions of the 2001 and 2007 reports are not that much different from each other. The 2007 report concludes in its 'Summary for policymakers', 'Most of the observed increase in global average temperatures since the mid-20th century is very likely due to the observed increase in anthropogenic GHG concentrations.' The authors had reduced the uncertainty a bit, but broadly speaking the conclusions were the same.[8]

So it's not clear when the fire alarm went off. It was ringing loudly for some time before anybody really noticed it.

Staying with the theme of increasing awareness, can you make your own risk assessment? What are the risks of *not acting* to mitigate climate and ecological changes? I mean, principally, what are the risks of continuing to burn fossil fuels and consume unsustainably? Do you now have an estimate of these risks, some embodied sense of them? Now compare these to the risks of *acting* to mitigate climate and ecological changes. There are risks on this side too. For instance, transitioning from a carbon economy and funding massive carbon sequestration projects *could* lead to recession. Reducing travel could lead to a loss of the richness of life and a sense of freedom. What's your risk assessment?

Would it help to compare these risks with the threat of the Covid-19 pandemic? Both threats are 'invisible'; we need to have the situation interpreted and explained. One difficulty is that we don't have the number of prominent people talking about how scary climate change is, as we did for Covid-19. Another is that when we're assessing threats, we still rely on a sense of *proximity*. Is that tiger coming closer or moving away? We need a story, a timescale of the threat getting closer.

In general, we humans are not very good at dealing with threats that seem distant, perhaps even many years into the future. What's changed in recent years is that extreme weather, fires, droughts, and floods have persuaded many people that a climate and ecological emergency is not just a distant phenomenon, it is upon us. We're a very visual species, so it helps to see graphic images of extreme weather events on the news.

How does one burning house relate to the other? The burning house of the climate and ecological emergency sits inside the cosmic burning house of samsara, like Russian dolls. Even if we succeed in averting the current climate and ecological emergency, we will still find ourselves in the burning house of samsara. At the same time, might they be one and the same burning house? Could we look at the climate and ecological emergency as an integral aspect of samsara? 'What's in the way is The Way', as my Indian mobile used to remind me!

If you follow the Buddha's teachings, climate action is an integral part of Dharma practice. Here's the lowdown: circumstances are way beyond our control. We need to respond to them as creatively as we can. There's no doubt that this will challenge our fear of impermanence and our desire to hold on and be in control. No one gets to sit this one out.

When we're considering existential threats, one of the first things that goes out of the window is our sense of humour. So in the next chapter we're doing climate comedy!

But before we go there, I invite you to take a moment to dwell in the heart.

Guided reflection: dwelling in the heart
Go to https://www.windhorsepublications.com/audio-resources/audio-resources-for-tbh/

- Purpose of this activity: to explore what's alive in you right now, in relation to the climate and ecological emergency, whatever it is: fear, excitement, distress, curiosity. To dwell in the heart, experiencing its fullness and emptiness.
- Time: 15 minutes.

8

Climate comedy

> I've been warned that telling people to panic about the climate crisis is a very dangerous thing to do. But don't worry. It's fine. Trust me, I've done this before and I can assure you it doesn't lead to anything.
>
> – Greta Thunberg[1]

When Greta Thunberg started her School Strike for Climate in August 2018, people were struck by how little she smiled. Eighteen months later she was still known for her blunt observations about the adults around her. However, she seemed to smile more, and to be more comfortable with media attention. The quote above, from Davos in January 2020, suggests that she's even started to use irony to get her point across.

As mentioned at the end of the previous chapter, when we're dealing with existential threats it's very easy to lose our sense of humour. This in itself can be dangerous, as irony in particular and humour in general can connect us to a bigger, alternative perspective. Under dictatorships, opposition survives in humour, irony, and satire. Even from a strictly

Dharmic perspective, a sense of humour can help to bring out the incongruity of the ego claiming anything at all for itself.[2] When we're talking about the climate and ecological emergency, a sense of humour can connect us to a bigger perspective. We need to find ways to voice our fears, and talking about them allows things to settle on a deeper level.

Bhikkhu Anālayo, a Buddhist teacher who presents an online course on Mindfully Facing Climate Change,[3] is also not known for his humour. However, he recounts a visit to his dentist. There he was in the dentist's chair with his shaven head and full Theravadin monk's robes. As the dentist leaned over Anālayo to begin the procedure, he said, 'You tell me if there is anything that bothers you.' Anālayo replied, 'Yeah, there is something that bothers me. It's climate change.' Everyone in the room laughed. Anālayo checked if there was some openness to hear more. Afterwards, the dentist came up to him and asked him for some reading.

What's funny about ecological Armageddon? I mean, you tell me. When I do this as a stand-up comedy, I ask the same question. We have to go right back to childhood, to when we laughed about ghosts and ghouls.

> Me: Knock knock.
> Audience: Who's there?
> Me: Ecological Armageddon.
> Audience: Ecological Armageddon who?
> Me: Ecological Armageddon outta here! (*I make as if to leave the stage.*)

Worth a groan? Let it out, I say. Better out than in, as my yoga teacher used to say.

Ecological Armageddon is no laughing matter. This puts me in a very awkward situation.

I'm reminded of my friend Suvarnaprabha. In one of her blogs she described a tour around a recycling plant in the Bay Area near where she lived in San Francisco. She was fascinated by what happened to the paper, metal, plastic, glass, and so on. Nothing disappeared, it all had an onward journey. Later, when she announced to the world that she had terminal cancer, in my mind this tour round the recycling plant became part of her preparation for death. Her blog concluded, 'Thinking of throwing something away? There is no away.' That's profound.

It's hard doing climate comedy. Just like the climate emergency, you're faced with impossible odds. Saving the planet is a lost cause. Did you hear the one about the small group of people that saved the world? No, I didn't either. And that leads me to my point, ladies and gentlemen (and those who don't identify as such), if we want to campaign to change the world nonviolently, research suggests that we will succeed with the active support of 3.5 per cent of the population. Many such campaigns have succeeded with less.[4]

But maybe I should just drop the whole evidence-based thing. Last week somebody accused me of recycling jokes! I mean, I tell you. Instead of throwing them away! This isn't the first time that I've dabbled in joke recycling. In 1992, when recycling was first all the rage, I created a joke recycling point at the local Green Fair. I was there next to the paper, metal, and glass recycling bins with a table and a sign that said Joke Recycling. People came and recycled their jokes, and I asked them to make a contribution to the Karuna Trust.[5]

Ten years later I invented the world's first joke recycling machine. I made it from a large cardboard box. You wrote out your joke on a special slip of paper and put it in an envelope in a hole in the top of the box. Then you turned a handle on the front of the box and a recycled joke came out of a tray below. As you can see, it used the most modern technology available at the time.

Of course, nearly all of those jokes were composted and went back to the earth long ago. However, one of them is still in circulation. I'll tell you exactly how it happened. It was a town fair on Parker's Piece in the centre of Cambridge, UK. I'd teamed up with Björn Hassler – do you remember him from chapter 2? – to have a stall. We put our table next to the recycling bins. At one point, a man in his early twenties with glasses and a beard approached.

He said, 'I've got a joke for you.' I said, 'Give us the 50 pence first, that's how it works.' He paid me then said, 'What's the difference between a fish?' It was Cambridge, after all! I said to him, 'Look mate, I'm sorry but you're going to have to give us another 50 pence if you want us to recycle that joke.' Nowadays we can be clever and say something like, 'One of its fins is both the same' or 'It's a question of scale(s)', but at the time it was simply a joke without a punchline, and therefore almost impossible to recycle.

Joke recycling aside, have you found it difficult to introduce the climate and ecological emergency into everyday conversation? Here's how it usually goes. Your friend says, 'I'm going to pick up my kids from school, and I'm taking the car.' You say, 'The climate is going to Hell in a handcart, and you're driving a mile to pick up the kids?' And your friend says, 'I'm going to pick up the kids, alright? It might rain.' And that's usually the end of the conversation.

Actually, I did find a way to bring the climate and ecological emergency into everyday conversation. It happened quite by accident. The same friend asked me, 'Where are you going for your holidays?' In a moment of inspiration, I said, 'We're going to Antarctica. Gesine and I are going to see the penguins in Antarctica.' She said, 'Wow, that's pretty cool!' I said, 'Not as cool as it used to be!'

Now you can take that joke on different levels. Going to see the penguins in Antarctica is not as cool as it used to be because of the air miles it takes to get there, and an awareness of how air pollution is contributing towards climate change. Secondly, it's not as cool as it used to be because average surface temperatures in the Antarctic and the Arctic have risen in recent years.

Sometimes, as a species, we act as if the earth belongs to us, for our use and pleasure. However, as we saw in chapter 3, it's the other way round. The earth doesn't belong to us; we belong to the earth. Why this should be funny, I just don't understand. We are entirely dependent on the earth for our survival. Our bodies are made of earth, water, fire, and air. We're just borrowing these elements. One day we will give them back to the earth, to be recycled. That is (if we're buried), if we haven't eaten so many preservatives that our bodies never break down!

Someone asked me to include at least one good belly laugh in this chapter. I looked down at my stomach, but I couldn't find a paunch-line! They told me that it was a bit below the belt. I still couldn't find one.

Buddhists are the ultimate recyclers. It's written into our tradition. We have to do it. We've got karma: what goes around comes around. And rebirth: because some jokes just. Keep. Coming. Back.

That's probably my funniest joke, apart from the one about the penguins. Maybe I should stop here. But the show must go on, so I caught up with Shantigarbha at his home in Bristol to ask him about joke recycling.

He looked at me with a knowing look and slurped his plum and ginger kombucha in a deliberate way. It was so deliberate that I thought he must have meant something by it. It's these small things that reveal the soul. And then he licked his plate, where there had been a coconut macaroon.

'Shantigarbha!' I exclaimed, 'You're embarrassing me. You're not at home now.'

'Oh, but I am!' he replied. I tried to switch the conversation back to the climate and ecological emergency: 'So tell me, how can climate comedy help us to come to terms with the climate and ecological emergency?'

He smiled and said prophetically, 'What will it benefit someone if they successfully address the climate and ecological emergency, yet lose their sense of humour?'

It was all getting rather self-referential, so I decided to consult a professional. As it happens, Tejopala, who appeared in chapter 5, has had a career in stand-up, and more recently, climate comedy. I caught up with him for my own personal climate comedy masterclass.

He began with, 'So, you want to steal the whole of my ninety-minute professional stand-up comedy routine, which took years to develop, for your book, do you?'

I replied, 'No, just the funny bits.'

He glared at me from the other side of the world.

I decided to switch tack: 'So tell me, when did you first see your future in climate comedy?'

'Back in the late nineties I was doing stand-up comedy. At the same time, I was just starting to get interested in the Dharma in a serious way. One night I found myself on a stage in Christchurch, NZ, and mid-routine I felt a strange tension between being authentic and being funny. I was trying to do both, but I felt like I had to choose one, and I chose being funny. When I got off the stage I realized that I couldn't do this anymore. It probably took me a year to stop. I kind of walked away. Then I was busy with going to the UK and getting ordained and so on. In 2009 a friend recommended that I do a course with a clown teacher called Giovanni Fusetti.[6]

'Giovanni teaches you how to play on your own physical and verbal quirks and tics, and the things that you usually try to hide about yourself. This creates a genuine and authentic character that you can always play with. So *whatever* you say out of that character is always authentic.'

He continued, 'It felt like I had a way back in, so I went and studied with him. What I really wanted to do was create a full-length clown show about the climate, with different scenes. I still want to do it. But what's much easier for me to do is stand-up, because I can do it on my own.

'The same friend also put me in touch with Bellina Raffy, who runs a company called Sustainable Stand-up.[7] Her thing is how to make warm, appreciative comedy, in a way that makes a difference in the world, particularly in relation to the environment. I ended up doing a show at the 2019 Melbourne

Fringe Festival, called Killing the Planet is against My Religion, which was about my life as a multifaith climate activist. And it worked and it didn't. It sold out, a tiny thirty-five-seater for six nights. It got a lot of laughs. But on the level of how authentic I was able to be, I still feel like there's quite a long way to go.

'I think that when you're writing comedy, if you reach for the thing that sounds funny first off, then you end up writing pretty mediocre comedy.'

'That's me out of the game, then,' I said.

'But if you try to write something meaningful – I don't mean bash people over the head, but something authentic – I've found that's probably my better material. Because stand-up relies on a premise and punchline. The premise itself is never funny. What you do with the premise is create a tension, and then you pop the tension with the punchline. You might do some other stuff like act from the character, but essentially those are the two main elements.'

He warmed to his theme: 'The tension exists in subjects that are not inherently funny. People ask, how do you write comedy about the climate? Well actually, there's a *hell* of a lot of tension involved in it. It's tough, but if you can find your way in, it's enormous. A lot of it depends on having enough emotional positivity to be able to laugh at it, in a way that isn't bleak.

'There are a few pitfalls. Do you laugh at it in a way that just leaves people in a cynical place? Or do you bash them over the head, just giving information? You don't really want to do that either.

'That's the theory. But the most common opportunity is an open-mic night, where people have just come to listen to comedy. People want to lynch me the moment I say that I'm a climate activist, because the most common response is, "Oh god, don't preach at us!"

'So I start by saying, "Hi, I'm Tejopala. I'm a climate activist." Then I wait for a response. If people cheer, I say, "That's interesting, because I usually get one of two responses. Either 'I hate you and I want to kill you' or 'Oh my god, you're going to bore me to death.' I kind of understand the second one, because I *have* trained with Al Gore in doing his full-length slideshow presentation. There are 538 slides in that show. By the time you're half-way through, you're thinking, I just wish the end of the world would hurry up. Come here, polar bear, I'm going to shoot you myself."

'Another time I might tell them that I'm a multifaith climate activist, but I'm also a stand-up comedian. That means that sometimes I end up trying to be funny in *really* inappropriate situations. When a group of us went in to meet Josh Frydenberg, the Minister for the Environment, to try to stop the Adani coal mine, I realized that we really were a bishop, a rabbi, a nun, a United Church minister, and me, an ordained Buddhist. And I'm turning to the others as we go into his office, saying, "Hey guys, get this . . .!"

'Then another thing I do is to lead them through a climate meditation. I say, in a soothing tone, "Now that you know that climate change is real, and there's no future, you can just relax and let go of the weighty moral responsibility of doing anything at all. Listen to the sound of a glacier . . . just cracking and floating away. And the sound of a baby polar bear [plaintive little cries] calling for her mother, who isn't there."

'Comedy, and humour generally, pops a tension. From a Dharmic perspective, ultimately you're trying to use humour to pop the tension of the deluded sense of self.[8] In the middle of the Vajrasattva mantra [a Buddhist mantra evoking primordial purity], I believe that's what that "Ha ha ha ha ho!" is about. I think that's a realization through humour that it was completely absurd to even impute a sense of self at all.'

I thanked Tejopala for articulating what I couldn't myself and went into a period of deep reflection. The words of Karl Marx floated through my mind. As a student of the broad sweep of history, Marx reminded us that history repeats itself, first as tragedy, second as farce. My guess is that if he had survived until the current epoch, he would have perceived a third possibility, tragi-comedy.

Mar-a-Lago: a tragi-comedy

Dramatis Personae

Donald Trump: POTUS
Greta Thunberg a.k.a. the goddess Athena
Chorus: 11,000 of the world's scientists
Cable News anchor

Act 1. November 2016, on TV

Cable News anchor: Tonight the nation is adjusting to the shock news that Donald Trump has beaten Hillary Clinton to the White House. His promise to Make America Great Again has obviously paid off. We'll go live to NYC where Donald Trump, the newly elected President of the United States, is giving a victory speech to his supporters.

Trump: As I said during the campaign, we're going to drain the swamp. . . . Working together, we will begin the urgent task of rebuilding our nation and renewing the American dream. . . . We have a great economic plan. We will double our growth and have the strongest economy anywhere in the world. . . . We must reclaim our country's destiny and dream big and bold and daring. (*Donald Trump smiles and combs his hair back into place. The camera cuts to an adoring crowd of Trump supporters waving a sea of flags.*)

Act 2. January 2019, on Twitter

Trump: Large parts of the country are suffering from tremendous amounts of snow and near record-setting cold. Amazing how big this system is. Wouldn't be bad to have a little of that good old-fashioned global warming right now! Still, I'm off to Mar-a-Lago to play golf! Plenty warm enough down there. I mean, shirtsleeves.

Greta (*replying on Twitter*): I want you to act as if your house was on fire.

Trump: She seems like a very happy young girl looking forward to a bright and wonderful future. So nice to see!

Greta: Listen to the science. It's not *opinion*. It's not *my* opinion. It's fact. You have stolen my dreams and my childhood with your empty words, yet I'm one of the lucky ones. People are suffering, people are dying, entire ecosystems are collapsing. We are in the beginning of a mass extinction and all you can talk about is money and fairy tales of eternal economic growth. How dare you. The eyes of all future generations are upon you, and if you choose to fail us, I say, we will never forgive you!

Trump: So ridiculous. Greta must work on her anger management problem, then go to a good old-fashioned movie with a friend! Chill Greta, chill!

Chorus of scientists (*via grainy video link, dressed in lab coats, dancing slowly in formation as they intone*):

Rivers are poisoned,
Fish leap no more.
CO_2 *levels are increasing,*
Coral reefs are no more.
The ice caps are melting,
Polar bears are no more.
Sea levels are rising,
Islands are no more.

(*Building to a frenzy.*)

What have we done?
We have hurt our mother.
The earth is our mother.
We have changed the weather.
We have hurt the earth.
We are acting like gods.
It's only the gods that
can change the weather.

(*More frenzied still.*)

The earth is warming.
They won't allow it.
The poles are melting.
They won't allow it.
The seas are rising.
They won't allow it.
They won't allow us
to rival the gods.

(*The chorus collapses into a wailing heap.*)

Act 3. Mar-a-Lago, Trump's 'Southern White House' in Florida

The Trump family are hosting President Bolsonaro of Brazil at a banquet in the Mar-a-Lago ballroom. The men gleam in white ties and tails. Tiaras on the heads of the women flash and sparkle in the ballroom lights.

Greta Thunberg appears dressed as the goddess Athena in combat slacks with helmet and semi-automatic assault weapon, hovering in the air above the throng. The crowd are awestruck and go silent. Donald Trump deliberately tries to ignore her and continues to talk to President Bolsonaro. Bolsonaro looks up at Greta.

Greta: Not you! I'll deal with you later. I've come for the balding, narcissistic buffoon. (*Trump slowly looks up.*) Yes, you! The gods have had enough of your trumpery, little man! You didn't listen to the science. Now, I'm working on my anger management, so our punishment will be somewhat humorous. First, you shall not win a second term.

Trump: Those mail-in ballots. I knew it!

Greta: Second, as sea levels rise, your Florida golf courses will be flooded.

Trump: No! It can't be! Security, where's security? How did she get in?

Greta: Third, I'm now performing cosmetic surgery to make you look like a Mexican greenkeeper. It shouldn't be difficult.

Trump: No! (*He panics as his skin changes colour.*)

Greta: Your fate is to tend the greens on your golf courses while WASPs play. Begone!

Trump: My golf courses! My golf courses! (*Trump's security staff see him as a Mexican, drag him from the table, wrestle him to the ground, handcuff him, and take him away in an electric golf cart.*)

Chorus of scientists (*from outside the ballroom*):

Life is long and
the unexpected comes to pass.
He thought he was equal to the gods.
He changed the weather.
What a fool he was!
Only the gods can control the weather.
He had to be punished.
Still, it could have been worse.

Those ancient Greeks were a punitive lot! If you committed *hubris* (violence, insolence) against the gods, you were in for some kind of vicious comeuppance.

If that's not enough for you, I've got an idea for a TV show. Shall we do a trial episode now? It's called – yes you guessed it – *Going to Hell in a Handcart*. It starts with the theme tune, which has a slowly descending melody. Think of Roberta Flack's 'Killing me softly with his song' combined with 'The best you can do' by Christopher Cross:

Yes, we're going to Hell in a handcart, baby,
Going to Hell in a handcart, baby,
Going to Hell in a handcart, baby,
Going to Hell!

'Hello I'm Shantigarbha, and welcome to *Going to Hell in a Handcart*. This is the show where we realistically try to imagine what the coming ecological Armageddon will be like.'

At this point I approach people in the audience holding an object that could be mistaken for a microphone, saying, 'You madam, what do you imagine the coming ecological Armageddon is going to be like?' Then I hold the microphone object out for them to speak. It always works.

'I can see floods and fires and food shortages. The destruction of the earth.'

'So a full-on biblical Apocalypse.' I move to the next person, 'And you sir, what do you see coming?'

'Nothing. I can't see anything. The window must be misted up.'

I reply, 'I think you're in the wrong movie. It's not a train. We're on a handcart, or pushing it for someone else. There are no windows.' I move on, 'And you madam, what have you got for us?'

'I see rainbows and unicorns!'

I say, 'Some people! Doesn't matter what you tell 'em.'

Wheeling away from the audience, I announce, 'That's all we've got time for this week! Thank you for joining us, and remember that a new world is just waiting to be born. On a quiet day, I can hear it breathing.'

Theme tune. Credits roll to audience applause.

In the next chapter, we're going to deal with all that anger that you've been experiencing. In the meantime I suggest taking ten minutes to soothe your heart by listening to birdsong.

Guided reflection: listening to birdsong[9]
Go to https://www.windhorsepublications.com/audio-resources/audio-resources-for-tbh/

- Purpose of this activity: to nourish and soothe your heart.
- Time: 10 minutes.

9

Transforming anger

> This is a dark time, filled with suffering and uncertainty. Like living cells in a larger body, it is natural that we feel the trauma of our world. So don't be afraid of the anguish you feel, or the anger or fear, because these responses arise from the depth of your caring and the truth of your interconnectedness with all beings.
>
> – Joanna Macy[1]

Anger and hatred are very live issues in environmental activism. Like Greta Thunberg's, my anger is usually directed at politicians, whom I judge to be 'doing nothing', oil and coal companies, loggers, US citizens (as having the largest average national carbon footprint), and other wealthy people. In one sense this anger is a distraction. I am no longer putting my attention on those who are suffering and how to alleviate their pain. Instead, I am caught up in how to punish those I believe to be responsible.

Buddhaghosa, the famous fifth-century Buddhist commentator, compared anger to a person picking up a burning ember or excrement to hit another. Whether or not they hit the other person, they first burn themselves or make themselves stink![2] So do Greta and I need anger management, as Donald Trump suggests? Is there really a problem with anger and hatred if they give us the energy to confront the causes of the problem? Isn't it better to get angry than to doze in our armchairs or privately seethe with resentment?

It's important to distinguish between anger and hatred. Anger can have a protective function or it may simply be a sudden release of frustrated energy. From a karmic perspective it may be neutral, just an intense feeling. You can feel anger in relation to someone without wishing to do them any harm. This intense energy can even be used skilfully. So anger is not entirely incompatible with *metta* (loving kindness), at least in the long run.

If I look at my own emotional history, there's more of a danger of getting stuck in repressing strong feelings such as anger and desire. Ironically, where I wasn't aware of what I was actually feeling, this unawareness *itself* was probably unskilful. In addition, repressing strong feelings meant that I was alienated from my energy. As we shall see later in the chapter, repressed anger tends to become guilt, shame, and depression. It's no coincidence that I suffered as a teenager and in my twenties from deep depressions. In this depressive state, experiencing my anger might actually have been a step forward.

Indeed, recontacting my emotions in any way I could was very important. As my Buddhist teacher, Urgyen Sangharakshita, said in an aphorism, 'Honest collision is better than dishonest collusion.'[3] However, even if anger is not intended to harm anyone, it can be very unpleasant to receive, and is likely to reduce trust and connection. So it's worth learning how to express it skilfully.

Hatred, on the other hand, is a desire for another person's suffering. It is by definition unwholesome and will lead to suffering for oneself and others. Hatred is likely to arise from frustrated craving. When we don't get what we want, we hate what's in the way, or even turn to hating the object of our craving. It can also arise from an unacknowledged lack of self-value, a fear of what others may think of us, or a failure to take responsibility for our suffering.

Through hatred, we become attached to our enemies just as strongly as if they were the objects of our craving. We simply can't do without them. We need regular 'fixes' of hatred, for instance by watching the news. According to the Dharma, hatred is the worst state of mind to get into, because it isolates us from others, thereby undermining our sense of interconnectedness.[4]

It's certainly less wholesome than craving. Craving shows an affinity with life, with other beings, and holds out the possibility of experiencing pleasure, however transient. Hatred is said to make us ugly – it certainly makes us unattractive to others. It even takes away our sleep. If we cultivate hatred, we are likely to become haunted by demons of destruction.

The best way to defuse hatred is to understand and appreciate the other person as a human being in the light of their own lives and conditions.[5] Who do you hate? Politicians, heads of multinational corporations, billionaire media moguls, climate-denying US presidents? Imagine their background, their history, all the things that made them who they are. The more you can do this, the more your energies will be directed not at the *person* but at their worldview and the ways they go about meeting their needs.

If you find yourself getting stuck with this, recall a time in your own history when you acted with a similar lack of awareness of others' needs. Did you change your behaviour because another person expressed their hatred towards you? Or was it because they talked to you as one human being to another? We can remind ourselves that people can and do change. As the Buddha famously said, 'Not by hatred will hatreds ever be pacified. They are pacified by love.'[6]

Nonviolent environmental activism can only be undermined by physical or verbal violence. Most importantly, violence won't work as a political strategy. As John Lennon remarked, 'When it gets down to having to use violence, then you are playing the system's game. The establishment will irritate you – pull your beard, flick your face – to make you fight. Because once they've got you violent, then they know how to handle you. The only thing they don't know how to handle is nonviolence and humour.'[7] We'll return to the theme of nonviolent social change in chapter 12.

Environmental activism works best when it rests on a broad societal consensus. If the public detects that an environmental activist movement is motivated by hatred, they won't support it. They know that political movements motivated by hatred, if they are successful, turn into authoritarian and repressive regimes. Hatred is ugly and repels the very people that environmental activists need to support them.

So let's not demonize anger. We needn't be afraid of getting angry, as long as we are careful not to allow it to develop into hatred. At the same time, we need to put effort into transforming our anger, expressing it and even receiving other people's anger in ways that increase safety, trust, and respect. Instead of repressing anger or 'managing' it, how do we find the life in it?

The key to this and most self-development lies in developing awareness. Awareness brings increased choice and freedom. If we become aware of the choices we're making as we act, this opens up the possibility of choosing differently in the future.

Looking into the experience of anger, we can see that it's strongly associated with 'should' thinking, thinking that implies wrongness. We have learned this kind of thinking in our families, at school, at work, even in our religions. Should thinking includes concepts of worth, deserving, reward, and punishment.

In Nonviolent Communication workshops, we say that there's nothing wrong with feeling anger and nothing wrong with the thoughts associated with it. If we say that anger is wrong, we fall back into the kind of thinking that created it in the first place. Paraphrasing Albert Einstein, 'We cannot solve our problems with the same thinking we used when we created them.'[8] So how do we approach anger with the intention of learning and healing, outside of the framework of right/wrong thinking? How *do* we 'find the life' in it?[9]

Let's take something specific and work with it. As you know, I find that photos of rotting carcasses of dead whales with kilogrammes of plastic in their stomachs are a particular trigger for me. The process for transforming anger that we follow in Nonviolent Communication suggests first to *stop and breathe*. As Marshall Rosenberg put it humorously, 'Count to a million slowly!'

The second step involves *getting in touch fully with what you are feeling, in your body. Where do you feel it?* In my case, I feel intense tension in my head and chest. I find it hard to breathe. At the same time I feel a kind of hopelessness in my heart, a desire to let it all go.

Third, *listen compassionately for the thoughts associated with these feelings*. I'm telling myself, 'We're killing the natural world.' 'What a horrible way to die, from starvation.' 'There are some real monsters out there, and they are human.' 'People just don't care.' 'If they could only see the consequences of their actions.'

Fourth, *get in touch with the need behind or underneath these thoughts. What are you really longing for in this situation?* I'm longing for care and protection for the natural world, especially for mammals that are self-conscious and similar to humans. I value reverence for life in all its forms. I'm longing for awareness of our collective actions, especially the impact of dumping plastic waste in the oceans.

Fifth, *get in touch with your 'transformed' feeling. Notice that when you are fully in touch with your need, the story drops away and you no longer feel angry – maybe sad, frustrated, in mourning.* I feel sadness, deep sadness. I'm mourning the loss of the sense of beauty and wonder that I experience when I see footage of whales swimming, fishing, or grazing. *If you feel a new surge of anger, go round again with the feeling, thoughts, and underlying needs.*

Sixth and finally, a step to complete and ground this process. *When you are settled in your transformed feeling, make a request to yourself to act on it right now. For instance, you might choose to express your feeling and need to another person, or find a way to meet your need in another context.* I choose to mention plastic in whales' stomachs when I'm launching this book, to keep it real, and bring in my own vulnerability.

So that was my transforming anger process. If you haven't already, I suggest that you repeat the process using your own example. For the sake of your emotional health and the planet, I recommend that you make a note of the request, and act on it!

Once we've 'found the life' in our own anger, we will need ways of receiving it in others. But first, why would you do that? Why would you stand there and receive another person's angry words? Perhaps you might do it to support a world where people connect and meet their needs peacefully. You might do it to contribute to the person in front of you. You might do it to support clarity and understanding. You might do it to care for the connection, or to try to see their humanity. You might do it in order to go for the heart connection and hang loose to the outcome.

This may be a particular challenge for people who are uncomfortable around anger. However, there's a lot going on in our world, and as I said above, there's nothing wrong with anger. If we want to be open to feedback, it's going to limit things to say, 'We'll only listen to you if you're under control, speaking rationally, and able to express yourself in a way that makes me comfortable.'

If someone is angry, that is part of what they need to communicate. To embody compassion, we could take it as our responsibility to learn how to receive that anger in a kindly manner. This means being kind to ourselves, knowing our boundaries and limitations. And it also means being able to hear where they are coming from. If they are coming from a very painful place, this means not expecting or requiring them to tone it down or talk calmly and politely.

If we're going to live from our inspirations, we need to be in touch with our passions. Sometimes our passions are unrefined and energetic, and can lead us to great places. We will not transform our anger by ignoring it. We will only transform our anger by working with it kindly, wisely, and boldly.[10]

Marshall Rosenberg pointed out that we have a choice about how we receive painful messages. As mentioned above, it seems to be a principle of self-development that awareness brings choice. So as we become more aware of the choices we're making about receiving anger, we increase our capacity to choose differently in the future.

In 2019 I joined the Nonviolence and De-escalation Team supporting Extinction Rebellion's Summer Uprising on the streets of Bristol. Despite Bristol City Council's best efforts to keep protestors and traffic apart, the 'Rebels' faced some angry motorists. One day I had just finished delivering a training session and went up Castle Park to see what was happening at one of the roadblocks.

I came across the following scene: four or five Rebels were blocking the road with a wide banner which read 'Act Now!' In front of them, a man in a van was revving up his engine, waving his hands and shouting at the Rebels, 'Get out of my way! Go and get a job!' One of the Rebels was saying, 'This is nothing compared to the disruption that's coming if we don't act now.' The van driver replied by pressing his horn and yelling, 'I've got deliveries to make. I've got a family at home to feed. You people are just causing trouble!'

As I was still wearing my white Nonviolence and De-escalation vest, I decided to approach the man, not head on, but from the side, which generally poses less of a threat. Broadly speaking there are four ways of receiving such a message as his. First, I could have chosen to receive it as blame and criticism and respond in a similar blaming manner. In a mild way, this is what the Rebels were doing when they were saying, 'This is nothing compared to the disruption that's coming if we don't act now.' My equivalent would have been, 'The earth is dying and you don't care! You're just increasing the pollution! You should start taking responsibility for your actions. You should get a different job that is less polluting.'

Second, I could have chosen to receive it as blame and criticism and blame myself. So in my head I might have said to myself, 'He's right. I've no right to be here, getting in people's way. It's my fault that he's angry. I should have thought about how people would react earlier. I should just get out of the way.'

Third, I could have chosen to take a moment for self-connection. Specifically, I could have sensed my feelings and needs. This is what I actually did. Internally, it sounded something like this, 'When I hear the engine being revved up, see him waving his arms through the windscreen, and hear him speaking at a volume that can be heard 10 metres away, I feel tense and a little bit scared. I want to know that I'm physically safe here. Underneath that, I'd like to be understood. Why am I supporting a group of people to stand in the street behind a banner saying Act Now! blocking the traffic? I'm longing for understanding for my terror of what we're doing to the planet, and what kind of planet we are leaving for future generations. I'd like the understanding that I'm doing this out of desperation, because I care for the earth.

Fourth, I could have chosen to empathize with the speaker. This is what I did next. I said, 'I can hear that you're angry because you want to be free to go about your business.'

The van driver replied, 'Yes, that's right!'

'And it's important to provide for your family, and this job is a way to do it?'

He responded, 'Yes. I've got enough to worry about without you lot blocking the streets.'

I continued empathizing, 'Things are stressful already without further delays.'

'Yeah. What's all this about anyway? Why do you have to do this?'

Here was my chance to tell him why I was there. So how do we express our anger fully? As I've said, anger isn't an undesirable quality that needs to be got rid of. How do we express the core of our anger fully and wholeheartedly? Other people's behaviour is the stimulus for our anger, not the cause of it. Other people don't *make* us feel anything. Our feelings come from our needs and longings. So expressing our anger fully comes from expressing the life in it, the underlying needs and longings.

So I told him, 'I'm terrified of what we're doing to the planet. What kind of planet are we leaving for future generations? I'm doing this out of desperation, because I care for the earth.'

He replied, 'That's all very well, but could you do this somewhere else? I've got to be back here tomorrow.'

I acknowledged him, 'So you don't mind us protesting, so long as we don't get in your way.'

He nodded, reversed his van, and went round another way.

Are you ready to start expressing your anger fully?

In the next chapter we're going to acknowledge our grief at the climate and ecological emergency. In the meantime I invite you to deepen your work in transforming anger. Part of the process of labelling and pigeonholing people is to create a hostile or enemy image of them. It's the mental equivalent of a visual caricature used for target practice! Relating to the image rather than the person makes it easier to dehumanize them, and so to justify violence towards them. Here is a guided reflection to help you transform these kinds of hostile or enemy images. Apart from reducing the violence in the world, this exercise allows us to reclaim a portion of our own human dignity.

Guided reflection: transforming enemy images[11]
Go to https://www.windhorsepublications.com/audio-resources/audio-resources-for-tbh/

- Purpose of this activity: to help you reconnect with your humanity by transforming hostile or enemy images.
- Time: 15 minutes.

10

Ecological grief

> The work of the mature person is to carry grief in one hand and gratitude in the other and to be stretched large by them. How much sorrow can I hold? That's how much gratitude I can give. If I carry only grief, I'll bend toward cynicism and despair. If I have only gratitude, I'll become saccharine and won't develop much compassion for other people's suffering. Grief keeps the heart fluid and soft, which helps make compassion possible.
>
> – Francis Weller[1]

It was the October 2019 XR Rebellion, outside the Home Office in London. The police had just come round telling each person that a Section 14 order had been put on the road. That meant that they could arrest anyone who stayed in the road without further warning.

Someone handed me the mic and I said to the crowd, 'I was going to offer a session on nonviolence and de-escalation, but now it doesn't seem like such a good idea to take people away from the road.' The guy looking after the mic whispered 'Good call', and I saw nods from a few seasoned campaigners.

'So instead, I'd like to offer a kind of regeneration thing. If you want to get comfortable to sit for about ten or fifteen minutes, I'll lead through a guided meditation.' People shifted to some hay bales on the left, and more people joined the group.

'I don't know about you, but I notice quite a bit of tension in my stomach and solar plexus. So let's start with the body. When you're ready, bring your attention to your skin, and then inside your skin, getting an overall sense of your body. And allow your attention to trickle down your torso, all the way down your legs, to your feet and toes. Picking up sensations in your toes, being farthest away from headquarters.

'Soles of the feet, heels and ankles, calves, knees . . . sitting bones, allowing the earth to take your weight . . . pelvis, lower back, upper back, spreading into the shoulder blades, and down the arms to the elbows, forearms, and hands, with fingers and thumbs. Sensing the life in your hands, warm or cool, tingly or numb. . . .

'Allowing the muscles at the back of the neck to relax and lengthen. . . . Scalp, ears, forehead, eyes, nose, mouth, jaw, throat, and chest, with heart and lungs. If it helps, following your breathing for eight or ten breaths. Now sensing what you're feeling, checking in around the heart. . . .

'And when you're ready, getting in touch with your intentions, for being here, for sitting in this road today outside the Home Office. Perhaps to do some healing, perhaps to help create the kind of world you want to live in – the kind of world that you'd like to pass on to your children and grandchildren.

'Inviting any figures of support or inspiration to support us today. Connecting to our ancestors.' I felt welcome tears behind my eyelids. 'Connecting to the people who would be here if they could, to those who support us to be here, to other species, to future generations. Inviting them all to be present here with us in this road today. Asking for their support and blessing.'

I chanted, 'May all beings be happy and well. May all beings be happy and well.' Others joined in and we chanted together for a while, then came to silence.

'When you're ready, in your own time, open your eyes and look around, taking in the other people who are here.' I paused. 'I'd like to hear from two or three people, what comes up for you?'

I handed the mic to a woman in her thirties and she said, 'I feel much more peaceful and centred than when I arrived this morning. And I feel sad.'

I asked, 'What's the sadness about?'

'I'm just sad about the planet we're leaving for our children.'

I said, 'A note of grief. I wonder who else feels this grief about the planet that we're handing on?' Nearly everyone's hand went up. 'It's worth giving this grief some space, some care. Space for mourning. It will give us some weight and solidity as we go through the day.' I paused. 'I guess that's it from me.'

People applauded. I got up from the ground, wondering about saying my name. But it didn't seem needed, so I walked away feeling vulnerable, and at the same time deeply fulfilled. It was as if I'd dropped a depth charge into my psyche. Writing this now, I can still feel the tearfulness and vulnerability.

In this chapter I'd like to explore the grief we may feel to witness the loss of species, of ecosystems, and natural beauty. It's worth looking at, because anxiety, trauma, overwhelm, and a sense of powerlessness may lead us to denial and block our ability to take action.

To help explore this, I talked to Rowan Tilly, who is a founding member of XR's Embedding Nonviolence Circle and has served on XR's Strategy Assembly. She's also asked for ordination into the Triratna Buddhist Order. I first noticed Rowan's affinity with grief in her email signature:

> Magnanimous Despair alone
> Could show me so divine a thing
> Where feeble Hope could ne'er have flown,
> But vainly flapp'd its tinsel wing.[2]

When we met on Zoom, I told her that I loved the idea of magnanimous despair: 'There's a kind of generosity to it.' Then I asked her how ecological grief had come into her life.

She replied, 'I spent most of my life engaged with it in one way or another. I always hoped that we could turn around this great big oil tanker, so to speak. Gradually, in the last fifteen years, there was a feeling of sand running out between one's fingers, as the possibility to turn it around became less and less.

'And then in the autumn of 2018 it became, "Oh, it's starting. Here it is." Of course, scientifically you couldn't pinpoint the point of no going back. But I suppose in my own consciousness there was a turning point as I realized that it was starting to unravel, and really feeling into that. The grief became much more intense at that time, and I'd wake up with it. I usually spend forty-five minutes with a cup of tea doing *metta* practice, and grief is alive in my heart.'

Maybe we need to mourn. Things might not be as easy for us in the future. We might lose aspects of our freedom, ease, and comfort. We've had a foretaste of this with Covid-19 lockdowns and travel restrictions. Unless we live in an equatorial region, other people and species in other countries are more likely to be impacted than us by rising sea levels, loss of forests, degradation of the soil, different weather patterns, and so on. However, we might sense the loss of beauty and richness in the natural world. We might sense the loss of care for future generations and a connection to our common humanity. Until we make space for this kind of mourning, we're unlikely to be able to comprehend the scale of the challenge and find the energy and motivation to address the practical issues that face us.

This is reminiscent of the grief work that Joanna Macy, Buddhist philosopher and activist, talks about.[3] She emphasizes the importance of acknowledging our grief for the world. She suggests simple rituals to do this. Imagine being in a crowd of people who have come to take part in the healing of our world. Stand opposite one of them. Recognize that they love this world. Feel your respect for them, bow towards them. You're looking into the eyes of someone who knows well the suffering of the world. They live with that knowledge and don't look away. They know that the Arctic ice is melting, and they haven't let it kill their love of life.

In another ritual, she suggests using symbolic objects to honour different aspects of our pain: dead leaves for sorrow and grief; a stone for fear; wood for anger in relation to injustice; and a bowl to symbolize despair. She claims that it's important to grieve, because it's a protest against the collective agreement to ignore the risks and carry on with business as usual. We tend to block the pain, afraid that it will overwhelm us. We suppress it *and* the wish for the preservation of life. We need to learn to face grief, fear, anger, and despair, and unblock fear of them. The world is trying to speak through us. When we are not afraid of these, nothing can stop us.

Zen master, poet, and peace activist Thich Nhat Hanh was once asked by his students what we need to do to save the world. They expected to hear his thoughts on the most effective forms of social and environmental action. However, he replied, 'What we most need to do is to hear within us the sounds of the Earth crying.'[4]

Grief is an aspect of compassion. It's part of how we respond to the suffering in the world. Can you hear the sounds of the earth crying? We are capable of resonating with the suffering of our world. Indeed, one of the ancient Indian words for compassion literally means 'trembling with'.[5] So we can reframe our pain for the world as compassion, an aspect of our interconnectedness with the world. Grief can show our love, fear can show courage, anger can show our longing for equity, and despair can give us the space for something new to arise. We can trust that love and pain are two sides of the same coin.

In chapter 1 we heard about Siddhartha, the Buddha-to-be, and his reaction to old age, sickness, and death. It hadn't sunk in that he was also subject to these existential risks. It was a shock when he found out, and he cut short his pleasure trip to go back to the palace. Back in the palace he lost his appetite for enjoyments of all kinds.

In a similar incident recounted in chapter 4 we saw how Siddhartha was riding in the country on a magnificent horse when he came across a field that was being ploughed. Seeing the torn grass, and the earth littered with dead worms and insects, he felt grief welling up inside him. It was as if his own relatives had been killed. Then he noticed the ploughman's sun-blackened skin and the oxen's distress. A deep sense of compassion arose in him, and he got down from his horse. Walking slowly over the ground he cried out, 'How wretched!'[6]

These two stories symbolize the birth of compassion. It's not an easy process, even for a Buddha-to-be. Siddhartha had to experience for himself the shock, grief, and even depression that comes from facing suffering, in order to discover true compassion. How could it be different for us?

Grief is an aspect of compassion, and compassion is an integral aspect of the birth of the Bodhisattva, the ideal Buddhist. It's the overwhelming upsurge of compassion, the longing to rescue all beings from the burning house of samsara, which makes them a Bodhisattva.[7] Tibetan Buddhists tell this story of Avalokiteshvara, their beloved Bodhisattva of compassion. Once upon a time he made a great vow to deliver all the people of Tibet from suffering and lead them to Enlightenment. He swore to keep his vow and made a wish that if he hesitated, his head would crack into ten pieces and his body would split into a thousand pieces.

He entered into a *samadhi*, a profound meditation on compassion, through which he aimed to lead all beings to Enlightenment. After a long period he emerged from his meditation to find that not even one hundredth of the beings of the Land of Snow had been helped by his practice. He was seized by a bitter sorrow and wondered, 'What is the use? I can do nothing for them. It is better for me to be happy and peaceful myself.' Immediately his head and body shattered into pieces. In his agony, he called out to Amitabha, his presiding Buddha, who came to his aid.

Amitabha blessed the shattered fragments and fashioned a new body with ten heads and a thousand arms, which could reach out in all directions. Then Amitabha set his own head on top of the new body. In this way, Avalokiteshvara was able to carry on his work of dispelling suffering far more effectively than before.[8] As one person, we can't save the whole world, or even the Land of Snow. However, if we join forces with others, so that we become many-headed and many-handed, we have greater capacities and reach. 'We' only need to be part of it.

This story can help us to understand our own grief and the dilemma we face when our desire to benefit all beings comes into contact with a profound realization of our limitations. When worlds collide, a new star may be born.

There's a growing body of research on ecological grief, or eco-anxiety. Indeed, trauma resulting from the climate and ecological emergency may be one of the biggest obstacles to collective action. Trauma can lead to a paralysing helplessness and may need a therapeutic response.

The scientists documenting the decline of the Great Barrier Reef in Australia during the marine heatwaves that bleached two thirds of it in 2016 and 2017 were among those who suffered in this way. More generally, it's important to recognize how the climate and ecological emergency can affect mental health. In 2018 this led social scientists from Australia and Canada to introduce the idea of ecological grief as an emotional side effect of environmental degradation. They argued that grief is a natural and legitimate response to ecological loss, and one that may become more common as climate impacts worsen.[9]

The American Psychological Association recognized the far-reaching tolls of the climate and ecological emergency on mental health in a 2017 report. It recommended steps to increase psychological resilience at both individual and community level.[10] Dr Patrick Kennedy-Williams, who set up a group called Climate Psychologists, specializes in climate anxiety. He points out that the cure to climate anxiety is the same as the cure for climate change – action. It is about getting out and doing something that helps. The group offers workshops for parents and others under the banner Anxiety to Action.[11]

So how do scientists, and the rest of us, protect our mental and emotional wellbeing in the face of these challenges? As Joanna Macy pointed out, we need a way to acknowledge and mourn our losses. Depression and the accompanying sense of paralysing helplessness arise when this mourning is incomplete. If we know how to mourn, we're less likely to get depressed. And if we do get depressed, it's likely to be less frequent, or lighter.

Mourning is a human need, just like celebrating life. In fact, they are two sides of the same coin. We mourn when life isn't going well, when needs aren't being met. We celebrate when life *is* going well, when needs *are* being met. As the quotation that heads this chapter says, we go forward with grief in one hand and gratitude in the other.

Moreover, mourning is a collective or communal need. It brings us in touch with our common humanity. Like celebration, it's best done collectively. In the West we've become private with mourning. Traditionally, and in other cultures, it's the realm of poetry, music, myth, symbol, and ritual. For the vast majority of human history we have lived in a community or village context. From the moment we are born we expect to be part of a group. Rituals such as funerals are important because they acknowledge grief and mourning, helping to honour our loss and bring out its collective aspect. Some traditional cultures give a year for bereavement, to 'live in the ashes'. In today's Western culture you're lucky if you get one week of 'compassionate leave'.

As the context of the climate and ecological emergency makes clear, mourning and grief are not just for the loss of a loved one. We can mourn losses of any kind: loves, hopes and dreams, capacities, species, ecosystems, and loss of trust in life. And like any other grief, ecological grief comes in waves. It's not a one-time event. When it comes it needs a safe space where all feelings are welcome. As Rumi says in his poem 'The guesthouse', 'Welcome and entertain them all.' We have forgotten that we're part of an ecosystem, part of a watershed. We're kin to all other animals. We need each other. It can be painful when we suddenly remember.

As with other kinds of grief, some things don't help. For instance, someone telling us, 'Perhaps it's time to move on?' or 'You'll get over it.' We need to learn how to mourn our losses. If we stay with this process, it can be healing and integrating. Indeed, if we go deeply into mourning, we find what is most precious in our lives. To fully inhabit this life, we first have to grieve all that we have lost.

Rowan Tilly again: 'How strange that grief can be so painful and yet so beautiful, because it's also a measure of our care. Then the grief becomes a source of energy that I can draw on. Some of the sense of beauty is the beauty of my own caring. The human capacity to care is also beautiful. When I reflect in this way, I realize that we're part of the universe, and the sacredness of it. We need some way of honouring the loss, honouring something that's much bigger than me or the story I tell myself. Then something comes alive in relation to it. This aliveness connects me to every other human who has lost something. There's something intensely humanizing about it. It becomes a source of strength and anchoring that enables me to do what I do in the world.'

What are the steps to this process of mourning and healing? It has to start with the story we're telling ourselves about what's happened. So much of our energy is caught up with this kind of thinking that we can't really start anywhere else. The story includes our judging and blaming – of ourselves and others. Make a note of all the 'shoulds' and 'shouldn'ts', for instance, 'They are destroying the planet. They should wake up before they wreck the earth for their children.'

At some point, we need to separate our thinking (the 'story') from what actually happened. So the second step is to recall what you actually saw and heard. In the case of the climate and ecological emergency, this may be diffuse and over a prolonged period, so I suggest recalling a specific instance, for instance hearing that the USA had left the Paris climate accord, or seeing news coverage of the 2019–2020 Australian bushfires. Stick to what you actually saw and heard, free from interpretations and evaluations, for instance, 'Last week I saw a photo in the news of a koala with white bandages on its paws.'

As you recall the specifics, sense how you feel now, in your body. This is the third step in the process. Grief isn't an abstraction, we need to encounter it physically, as a bodily experience. Depending on the intensity of the grief, you might be feeling heavy, sad, anxious, tight in the chest, or distressed. Make a note of it, mentally or on paper.

When you're ready, explore behind or underneath this feeling, in a bodily way. The fourth step explores the significance of what has happened. Why does it matter so much to you? What are you longing for? What do you need and value? Longing is our connection to life. Again, I suggest that you make a note of it.

When you are fully connected to your longing, sense again how you feel in your body. This fifth step is important in order to keep a somatic connection as we go through the process. If we have support and periods of genuine solitude, grief will transform itself into tender melancholy. This 'sweet' sadness feels very different from the earlier feelings, because we are now connected to our longings rather than our thoughts.

The sixth and final step is to complete and ground the process. Imagine one step that you can take to honour the loss. What could it be? Perhaps to talk about your grief to others. Perhaps to create a ritual or spot for remembering it. Perhaps to dedicate your activities in the next week to this grief. Whatever gives you a sense of honouring the loss in a congruent way.

If we want to care for the earth and protect future generations, let's fall in love with the earth all over again! We belong to the earth, not the other way round. This way we'll stay sensitive to the losses around us.

The Buddha spoke of going beyond grief and lamentation. What could he have meant by this? The closest I've found is a modern poem about a *dakini* – a fearless mythical sky dancer from the Indian and Tibetan Tantric tradition.

The Dakini Speaks

My friends, let's grow up.
Let's stop pretending we don't know the deal here.
Or if we truly haven't noticed, let's wake up and notice.
Look: Everything that can be lost, will be lost.
It's simple – how could we have missed it for so long?
Let's grieve our losses fully, like ripe human beings,
But please, let's not be so shocked by them.
Let's not act so betrayed,
As though life had broken her secret promise to us.
Impermanence is life's only promise to us,
And she keeps it with ruthless impeccability.
To a child she seems cruel, but she is only wild,
And her compassion exquisitely precise:
Brilliantly penetrating, luminous with truth,
She strips away the unreal to show us the real.
This is the true ride – let's give ourselves to it!
Let's stop making deals for a safe passage:
There isn't one anyway, and the cost is too high.
We are not children anymore.
The true human adult gives everything for what cannot
be lost.
Let's dance the wild dance of no hope!

– Jennifer Welwood[12]

For the guided reflection I invite you to go more deeply into the steps to mourning and healing that I outlined above.

Guided reflection: steps to mourning and healing (grief work)

Go to https://www.windhorsepublications.com/audio-resources/audio-resources-for-tbh/

- Purpose of this activity: to create a reflective space for grief and mourning.
- Time: 15 minutes.

11

Gratitude

The Peace of Wild Things

When despair for the world grows in me
and I wake in the night at the least sound
in fear of what my life and my children's lives may be,
I go and lie down where the wood drake
rests in his beauty on the water, and the great heron feeds.
I come into the peace of wild things
who do not tax their lives with forethought
of grief. I come into the presence of still water.
And I feel above me the day-blind stars
waiting with their light. For a time
I rest in the grace of the world, and am free.

– Wendell Berry[1]

In 2019, working with Extinction Rebellion, I coordinated nonviolence and de-escalation (N&D) training across all nine London sites of the October Rebellion. By Thursday of the first week the centre of activities had moved to Trafalgar Square. There were dozens of tents crowded onto the central island,

and hundreds of Rebels blocking the traffic on the south side. Coming onto the square and looking up at Nelson's Column, I saw that the plinth was skirted with vivid red flames.

Later that morning I facilitated an intro session for new Rebels in the N&D tent. Straight afterwards they were keen to try their new skills, so I took them out onto the road on the south side. We immediately came across a tense situation. A line of police officers were shouting, 'Move back onto the square!' to fifty to a hundred Rebels who were standing in the road. As they shouted to the Rebels, they pushed them against a larger crowd of Rebels sitting on the pavement. I was very concerned for the safety of those sitting down. The way things were going, the standing Rebels could easily topple onto them.

Surrounded by the new recruits, I started singing. My intention was to distract the Rebels' attention from the police and de-escalate things. I sang, 'People gonna rise like water, gonna turn this system round. In the voice of my great granddaughter, climate justice now!' My new recruits joined in with me, but none of the crowd in front of us. Their attention was firmly on resisting the police. And the attention of the police was firmly on pushing them off the road.

I felt frustrated and helpless. How to break this deadlock? Then I heard a voice from the other side of the sitting Rebels. Looking over, I could tell from the white vest that one of the N&D team had got hold of a megaphone. In measured tones, he said, 'I invite you to take a breath . . . and another . . . Feel your feet, the ground underneath your feet, your connection with the earth.'

From the voice and the glasses, I could now identify him as one of my colleagues, Ben Yeger. He continued,

'Take a moment to remember why you're here . . . what your intention is.'

As Ben was talking, I slowly picked my way round the crowd until I was by his side. After a long pause, he said, 'Reach out and touch the person next to you.' I could feel in my stomach the tension of the crowd reducing.

I touched Ben on the arm and whispered to him, 'I've got something for a few minutes, when you're done.'

He looked at me with total relief and immediately handed me the megaphone. As I took it I remember thinking, 'I'm not ready for this.' However, the megaphone was already in my hand, so I took a breath and said to the crowd, 'Now turn to your neighbour again, and tell them something that you're grateful for, being alive on this earth.' I paused. 'Be specific. If you love the smell of coffee in the morning, say that.'

Ben said, 'That's beautiful!' and then emphatically, 'Now stay here and hold the space for a while.' I nodded as he walked away.

So I just stood there, with the megaphone in my hand, surrounded by the two or three recruits who had followed me. Within a minute the crowd was buzzing with relaxed conversation. I looked over the crowd and saw that the police had stopped pushing the standing Rebels. It struck me that once the Rebels relaxed and stopped resisting, the police had nothing to push against. They simply couldn't carry on pushing.

A quiet, burbly folk duo on flute and guitar started up to the left of the crowd, signalling a further relaxation. A man next to me in his mid-twenties asked me what I was grateful for, being alive on this earth. I replied, without thinking,

'I'm grateful for the love of my partner, Gesine. And feeling the warmth of the sun on my skin.' We must have chatted for twenty minutes. By now it felt like a sociable Sunday afternoon. Finally, a samba band marched through, changing the mood of the crowd again. We looked at each other and decided that our job was done, so we left the crowd to it.

In chapter 3 we saw how the newly Enlightened Buddha looked back at the bodhi tree and felt deep gratitude for how it had sheltered him. This reminds us that gratitude is an Enlightened quality – knowing that something has been done for our benefit. We also heard of Maha Kassapa's love of the mountains, 'Strung with garlands of flowering vines, / This patch of earth delights the mind; / The lovely calls of elephants sound – / These rocky crags do please me so!'

But perhaps the most profound evocation of beauty and transience comes from the eighth- to ninth-century Japanese monk Kukai, also known as Kobo Daishi. He's replying to a nobleman who asks why Kukai prefers to live up a mountain and meditate rather than enjoy life in the city:

> Have you not seen, O have you not seen,
> The peach and plum blossoms in the royal garden?
> They must be in full bloom, pink and fragrant,
> Now opening in the April showers, now falling in the spring gales;
> Flying high and low, all over the garden the petals scatter.
> Some sprigs may be plucked by the strolling spring maidens,
> And the flying petals picked by the flittering spring orioles.

Have you not seen, O have you not seen,
The water gushing up in the divine spring of the garden?
No sooner does it arise than it flows away forever:
Thousands of shining lines flow as they come forth,
Flowing, flowing, flowing into an unfathomable abyss;
Turning, whirling again, they flow on forever,
And no one knows where they will stop.[2]

We need gratitude in our lives because it gives us energy for living. The ancients celebrated gratitude as an important virtue in ways that might seem quite foreign to our ears. For example, the Roman philosopher and politician Cicero said that being and appearing grateful is not only the greatest of all virtues but the mother of all others.[3] Marshall Rosenberg, founder of Nonviolent Communication, described gratitude as giraffe juice, or fuel to fill up your energy tanks. Deliberately fostering gratitude corresponds to the ancient practice of counting your blessings, celebrated in the Christian hymn: 'Count your many blessings – ev'ry doubt will fly, / And you will be singing as the days go by.'[4]

In the Buddhist tradition, we have the delightful story of the three Anuruddhas, also told in chapter 3. We saw there how they avoided polluting water sources, demonstrating an acute ecological awareness. In this chapter another aspect of how they live together becomes significant: they practise gratitude towards each other.

When the Buddha came to visit them, he said, 'I hope, Anuruddha, that you are all living in concord, with mutual appreciation, without disputing, blending like milk and water, viewing each other with kindly eyes.' They replied, yes. The Buddha was curious to know more, how did they do it? The answer was by cultivating gratitude towards each other. They reflected that it was a great benefit to live with such companions in the holy life. On the basis of this reflection, they directed loving kindness towards each other in body, speech, and mind.[5] Reflecting on benefits received stimulates gratitude, which in turn provides the energy for further beneficial actions. It's a 'virtuous' cycle.

My own introduction to gratitude as a practice came when I was in my forties. Before that I was dealing with too much anger and grief to be able to experience gratitude. This is why chapters 9–11 are ordered as anger, grief, and gratitude. In my life journey I've found that I've needed to transform anger and acknowledge grief before I can get to feelings of gratitude. However, it's been worth the wait. In 2008 I was leading a session on gratitude at an NVC training. At the end of the session I felt so tender, so human, that I resolved to keep a gratitude diary. Thirteen years later I'm still keeping it.

It began as a small pocketbook that I wrote in just before falling asleep. I wrote whatever I was grateful for during the day. Sometimes it was a real struggle to find something. Other times it came easily. Wherever I was in the world, I left the diary open beside the bed during the night. That was important. I told people that it was so that the gratitude could perfume the room during the night. And during the day I had a rule that I would never put an object on top of the diary, so that it always remained open to the air. In this way gratitude slowly became part of my life.

My partner, Gesine, and I have an agreement that we will

do gratitude at the same time, just before we go to sleep. The only difference is that we do it verbally, rather than in written form. Even if we've had an argument, or there is tension between us, we still try to do this practice. In fact it's one of our ways to reconnect. We tell each other what we enjoyed and what we feel grateful for during the day. Of course there's space to share things that we didn't enjoy as well, so there's space for mourning.

As you can imagine, it really brings us together when we express something that the other one did that we feel grateful for. To make it really juicy, it seems to help when we're specific about what the other person has done, and which needs were met. For the receiver of this type of gratitude, the joy is to hear specifically how they have contributed to your life. Here's an example: 'I'm grateful for the meal you prepared this evening. I particularly appreciated the extra effort you made to roast the vegetables. I love the texture and taste, and the energy they give me. I received it as care and love.' This much more satisfying to hear than, 'You're an angel.'

To strengthen your connection with the earth, you could record your gratitude for sunlight, falling leaves, and bird-song. What do these give us? A sense of the preciousness of life, and beauty.

By making a daily practice of expressing your gratitude in thought and action, you start to rewire your habits of thinking along more wholesome lines. It becomes a kind of rebalancing of your attention. We've been trained to focus our attention on what's wrong as a first step to fixing it. This is very sensible and practical, but it can give us a kind of 'negativity bias'. Such a bias had an evolutionary benefit. Our ancestors were surrounded by predators only too willing to deprive them of life. If one of our ancestors heard a twig snap and jumped or ran away nine times out of ten, they saved their life. But if on the tenth occasion they ignored the twig, and it turned out to be a tiger, then they would lose their life and the opportunity to procreate. In fact that wasn't *our* ancestor. Our ancestors were the ones who jumped or ran *every* time. That's how they survived and propagated their genes.

So we're naturally cautious and tend to emphasize what isn't working. Gratitude becomes a way of rebalancing our attention, being more accurate, whilst ensuring that we don't fall easy prey to tigers! This rebalancing is a gradual process. We can tune into a deeper sense of joy, abundance, and connection. When the heart fills up with gratitude, it ultimately spills over and flows into the world in the form of kindness, empathy, generosity, and compassion. When you're starting this practice, just notice what is there for you, rather than trying to force something to come. If mourning comes up, rather than pushing it away make space for it and acknowledge what you're longing for before going back to gratitude.

In recent years, positive psychology has grown in popularity. It focuses on happiness and its causes, rather than suffering and its causes. According to empirical studies, keeping a gratitude diary is an 'intervention' that promotes lasting happiness and decreases depressive symptoms. Gratitude makes you happier! Showing gratitude for met needs is the most powerful happiness-boosting activity there is.[6] A conscious focus on blessings has emotional and interpersonal benefits.[7] As Gesine and I have found, gratitude acts as a booster shot for romantic relationships.[8] It also generates the urge to reciprocate benefits received by making a gift to someone else. Such reciprocation strengthens social networks.[9]

As I mentioned at the start of the chapter, gratitude forms the primal movement of many spiritual and earth traditions. It is grounding and strengthening and gives us the sense that we belong here on this earth. We can tune into it at any time. Let your feet feel the solidity of the earth! Allow the earth to take your weight.

We can feel gratitude for the gift of life, which is both delicious and terrifying! We may come to recognize that everything is a gift. Gratitude is the root of *joie de vivre*. It's not joy that makes us grateful, it's gratitude that makes us joyful.

Gratitude features strongly in the first and second of the four reminders from the Tibetan Buddhist tradition. They are taken from the *lam rim* or 'stages of the path'. These reflections are placed at the start of the path, as a preparation for other practices. They even precede going for refuge because they provide reasons for committing to the Buddhist path. The four reminders are: the precious opportunity offered by human life; death and impermanence; actions having consequences; and that suffering is a part of life.[10]

The first reminder invites us to reflect on the preciousness of human life. According to the Buddhist scriptures, the chance of being born as a human being rather than some other form of life is infinitesimally small. The Buddha invited his followers to imagine that the whole earth is covered with water. A man throws a yoke with a single hole onto the waters, where it floats this way and that according to the four winds. Now suppose that there was a blind sea turtle swimming in those waters, and this turtle came to the surface once every hundred years. What are the chances of the blind sea turtle sticking its neck into the yoke? It would be sheer coincidence. The Buddha concluded, 'It's likewise a sheer coincidence that one attains the human state.'[11]

You can turn over this image in your mind, and make the following reflections, if they are true for you:[12]

- Here, now, I have the chance to make something of my life. I have health, I have energy. I have the ability to think and feel freely. I have enough food and enough money to meet my needs. I live in a country that is free of war and many other difficulties that people can face. I'm not trapped in states of mind such as madness, craving, hatred, or depression. All of these things can change, but while I have these advantages, I have a great opportunity.
- I have had the great good fortune to come across the Buddha, Dharma, and Sangha. These have made practice possible for me.
- Am I making use of the opportunity this offers? How much of my time I waste! How much of my life passes in unawareness! How strongly my habits constrain me! I would be foolish to waste this chance. So let me commit myself to practising as fully as I can.

The second reminder invites us to reflect on death. By mentally subtracting the blessing of life, we can more easily appreciate and feel gratitude for it.

- One day I will die. I cannot avoid it. It comes to everyone, and it will come to me. Everyone who has lived in the past has aged and died, and those living now are aging and will die too. However, the time of my death is uncertain. Even though I know I will die, I don't know when. I need to stay aware of this, and to make the most of my life's opportunity.

Gratitude is liberating and subversive because it contradicts the dominant message of consumer society that you are not enough. Hearing this message, people feel inferior and mistakenly believe that they need to acquire things in order to feel 'enough'. Gratitude helps people to realize that they are sufficient. Could it be a gateway to a sustainable relationship with the earth?

When I did the megaphone de-escalation in Trafalgar Square I was inspired by Joanna Macy's suggestions for a gratitude exercise.[13] She suggests completing open sentences, for instance, 'Some things I love about being on this earth are . . .' Do you remember my spontaneous and heartfelt answer was the love of my partner Gesine and the warmth of the sun on my skin? Others might be: 'A place that was magical or wonderful to me as a child was . . .' 'A person in my life who helped me believe in myself is/was . . .' 'Some things I appreciate about myself are . . .' I suggest that you take a moment now to complete these sentences and enjoy the gratitude that flows.

If you want to take the practice of gratitude further, here are some questions that you could consider:

- Who benefited your life? What do you want to say to them? Write it down, say it to them later, or make it the occasion for a gift to someone else.
- What is your most precious memory?
- Is there a writer, artist, actor, or director who inspires you?
- How have you grown emotionally and 'spiritually' in your life?
- What insight or understanding do you have that you always want to remember?
- What is the appreciation of gratitude that you have received that you feel happiest about?
- What did you smile about today?
- Who inspires you to fulfil your potential as a human being?
- When did nature stop you in your tracks with her beauty?
- Who has helped you develop your 'inner compass' – given you a sense of direction?
- What is the greatest gift that you have ever received?
- Who could you appreciate more fully?
- What skill or quality do you most value in yourself?
- What are you grateful for in this moment?
- What challenges have you faced and losses have you sustained that have given you deeper roots of self-compassion?

In the final three chapters we'll be turning towards nonviolent social change and the kinds of collective actions that could be proportionate to the climate and ecological emergency. As preparation I invite you to draw on the power of gratitude with a guided reflection on gratitude towards the earth.

Guided reflection: gratitude towards the earth
Go to https://www.windhorsepublications.com/audio-resources/audio-resources-for-tbh/

- Purpose of this activity: to draw on the power of gratitude.
- Time: 15 minutes.

12

Nonviolent social change

> My final words of advice to you are Educate, Agitate and Organize; have faith in yourself. With justice on our side I do not see how we can lose our battle. The battle to me is a matter of joy. The battle is in the fullest sense spiritual. There is nothing material or social in it. For ours is a battle not for wealth or for power. It is a battle for freedom.
>
> – Dr B.R. Ambedkar[1]

In the final three chapters we will consider what needs to happen next, especially collectively. To guide us we have the life-enriching ethical principle of nonviolence or love for the earth and each other. How do we transform society in a nonviolent manner towards greater social and climate justice? Faced with hierarchical systems that enshrine economic, political, social, and religious power, how do we show another way that is less costly and more satisfying?

By now we have found a greater perspective and healing through the topics of this book: acknowledging that we're in a climate and ecological crisis, strengthening our connection with the earth, understanding Buddhist environmental ethics, deepening compassion, applying the Buddhist precepts, recognizing that we need a cry of inspiration, using humour, finding the life in our anger, witnessing our grief, and expressing our gratitude.

Hopefully our energies are now freed up to explore the kind of social practices and institutions that might be in alignment with our values and proportionate to the climate and ecological emergency. By social practices I mean repeated interactions, which could include relatively informal ones like family meals to more formal ones like social clubs, banks, NGOs, even the international air traffic system.

To help us engage and think creatively at this level of social organization I'd like to tell a story that I got from Marshall Rosenberg, the developer of Nonviolent Communication. He was fond of telling it at the start of his workshops on social change.

Imagine that you live on the bank of a river. One day you're looking out over the river and you see something floating down the stream. You can't tell what it is, but it catches your eye enough to make you wade out into the shallows. From there you can see the head of a baby. Shocked, you splash out further and swim to where the baby is. You pull the baby towards you and swim back to the shore. On land you find a way to dry the baby and warm it up. Later you find a way to feed it.

The next day you're back on the bank of the river for your usual walk. You're thinking about what happened yesterday, casting an eye occasionally over the river. Suddenly you see a shape in the water, then another. Two babies. You wade and then swim out to them, tuck them into your chest, and bring them ashore. Now you have three babies to look after, so you ask your friends and neighbours for help.

On the third day you're standing on the riverbank, staring at the river. Suddenly there are four shapes in the water. You rescue two, then the other two. What to do with them? The next day, eight. The next day, sixteen. What happens when there are more babies in the water than you can rescue, even with help? At some point you may ask yourself, 'Would I be using my resources more effectively to go upstream and find out how the babies are ending up in the water?'

Take a moment to reflect on where you see yourself in this story. What do you have energy for? It's not an either/or situation. It could be both/and. Some people are happy to keep on rescuing babies. Some people would go upstream on the second day. We all have our part to play. I see myself as the person who calls a community gathering to elicit a collective response: 'Hey, this is happening. What are we going to do about it?' When we know what resources we've got, we can make a plan to address the situation.

I'd like to bring in Joanna Macy, Buddhist philosopher and eco-activist. She has 'gone upstream' and checked out what's leading to the climate and ecological emergency. In her famous book *Coming back to Life*[2] she refers to an earlier upstream explorer, the Norwegian eco-philosopher Sigmund Kvaløy. He characterized our current economic and social order as an Industrial Growth Society.[3] When we examine the details of our economy it turns out that it *literally* depends on an ever-increasing consumption of resources.

The earth has become our supply house and our sewer. Clearly it only has a finite capacity for both. Our demands on the earth are accelerating, as the Industrial Growth Society demands 'growth'. The financial system depends on credit and an assumption of growth to pay back loans. Faced with dire predictions about 2–4°C temperature rises, it's important to remember the possibility that we can still meet our needs without destroying our life-support system.

However, to achieve this will require what Joanna Macy calls the Great Turning. This is the transformation from an Industrial Growth Society to Life-Serving Systems. These sustain the needs of humans and other life forms without jeopardizing the prospects of future generations. Macy points out that this Great Turning is already happening, and it's happening in three essential areas: actions to slow the damage to the earth and its beings; analysis of structural causes and the creation of structural alternatives; and a fundamental shift in worldview and values.[4]

Let's look at holding actions. At their simplest, these are about rescuing babies. It's asking your friends and neighbours to help. It's setting up an NGO to rescue babies and find homes for them. It's going upstream and holding up 'Stop' signs to protest against babies being abandoned. It's campaigning for laws to mitigate the impact of the climate and ecological emergency and promote climate justice.

It's even using force, in a protective manner, to save the lives of babies. For me, using force to protect needs is consistent with the principle of nonviolence. The intention is important here: it's not to punish someone or teach them a lesson, simply to protect needs. My Buddhist teacher Urgyen Sangharakshita described this as the power mode in the service of the love mode. Nonviolent disruption clearly falls into this category, as it's disrupting business as usual. You may regard any kind of disruption as 'un-Buddhist' or even unskilful; however, we have a clear example from the Buddha's life. When the two sides of his family went to war, the Buddha stood between the battle lines.[5] What is that if not disrupting fifth-century BCE business as usual? We'll return to nonviolent disruption in chapter 14.

Holding actions buy time. They save some lives, ecosystems, species, and cultures for the future. In themselves holding actions are insufficient to bring about a change to Life-Serving Systems.

Systemic analysis is about going upstream to find out why babies are ending up in the river. It can be a long and arduous journey to do the studies necessary to establish probable causes. In order to free ourselves from the grip of the Industrial Growth Society we need to understand its dynamics. What are the agreements and causes that have led to ever-increasing inequality and crisis?

The Industrial Growth Society has prioritized economic value over other forms of value. And it has placed economic value on raw materials, for example, and not on the impact of growing these on the local environment. These impacts are not costed and therefore appear to have no value. Such an 'extractive' system is part of a broader system that justifies the use of 'power over' forms of control.

Here are some of the characteristics of social systems based on 'power over': they sanction punishment and reward, shame and guilt, duty and obligation. They make violence against 'enemies' of the group enjoyable. They establish education systems to teach obedience to authority. Most importantly, they sanction 'prejudicial thinking'. This is the kind of thinking that values people differently based on the colour of their skin, caste, social class, race, religion, gender identity, sexual orientation, and physical ability. It is also the kind of thinking that leads to devaluing other species, ecosystems, and the wellbeing of future generations.

The world isn't a level playing field. Resources are distributed inequitably according to these different kinds of prejudicial thinking. As a follower of the Buddha I commit myself to actions to ensure a more equitable distribution of resources worldwide. Whether it's social justice more broadly or climate justice in particular, we need to set up the structural alternatives of Life-Serving Systems.

Such systems are based on the understanding that we and all living creatures, now and in the future, have needs. Secondly, we have a feedback system, called feelings or experiences. These tell us when needs are met or unmet and they mobilize energy to meet needs. Life-Serving Systems train us to think in terms of needs, feelings, and strategies to meet needs.

We need to learn about and support new structural shoots in economics and resource allocation, decision-making, justice, education, healthcare, energy production, animal farming, and so on. As one of the twentieth century's geniuses, R. Buckminster Fuller, observed, you can never change things by fighting the existing reality. To change something build a new model that makes the existing model obsolete. More on these new shoots in the next chapter.

Holding actions and a systemic analysis can't take root and survive without deeply held values to sustain them.

Such actions need to reflect how we relate to the earth and each other. They require a profound shift in our perception of reality. So far this book has been mainly about this kind of shift in consciousness, from a Buddhist perspective. The insights have been gained by reflecting on experience, seeing the bigger picture of the interconnectedness of life, finding beauty in the world and in people around us, and transforming our grief and anger into life-serving energy. Through these insights we can awaken to what we once knew, that the natural world is alive, full of life that resonates with our own lives, and is as valuable as life.

These insights will free us from the straitjacket of the Industrial Growth Society, giving us deeper goals and satisfactions. They will help us to redefine our understanding of wealth and sense of worth. They will liberate us from the illusion that owning things will bring us happiness. They will take us beyond the constriction of a belief in a separate self to a deeper sense of interconnectedness with each other, and a profound sense of belonging to the body of the earth.

We can find such a shift in consciousness in Engaged Buddhism, in the emerging Ecodharma and Ecosattva initiatives, and similar movements in other religions.[6] We can take heart that many are embracing a life of relative simplicity, by reflecting on what they really need for their wellbeing. Paraphrasing Gandhi, they are living simply so that others may simply live. We can recognize that nobody's needs get met unless everybody's needs get met. My need for food and sustenance is only partially met when my belly is full. It is fully met when all humans have food in their bellies. And it's hard to separate meeting human needs from the needs of the environment. They are one and the same in an interdependent scheme of life.

We can take courage and inspiration from the Buddha's commitment to nonviolent social change. He challenged the domination systems of his time and the prejudicial thinking that underpinned them. He undermined the claim of Brahmanical superiority, arguing that it's your worth as a person rather than your birth (family background) that counts.[7] And he rejected the widespread Brahmanical practice of blood sacrifice, with some evidence that this brought about a societal change in attitudes towards animals.[8]

Two hundred years later the emperor Ashoka built on the successes of his grandfather to establish a vast Mauryan empire stretching from the Persian border to the Ganges delta. Afterwards he was touched by the cost in blood and suffering, and he turned towards Buddhism. Instead of fighting aggressive wars to acquire new territories, he reimagined the role of emperor as protector of his people and, gradually, of all living beings in his empire.

On rocks and pillars throughout the empire he had carved the message of nonviolence and non-destructiveness. The seventh Pillar Edict declares that abstaining from injuring and killing living beings is the way that his subjects can demonstrate their devotion to the Dharma. He even took the first precept personally, forbidding the killing of animals in the royal kitchens, except for a few swans for the royal table![9]

In twentieth-century India two figures stand out in the history of nonviolent social change. Mohandas Gandhi was born and died a Hindu. Gandhi turned nonviolent civil disobedience into a powerful political tactic against the British administration in the struggle for independence. We'll look more closely at his famous Salt March in chapter 14.

By contrast, Dr Bhimrao Ambedkar began life as an 'untouchable' and died a Buddhist. He represents something absolutely unique in modern Buddhism. He overcame immense odds to fulfil his personal potential and earned a place in history by drafting the Indian constitution. However, his place here rests on nonviolently leading hundreds of thousands of 'untouchables' from the hell of caste to the dignity of Buddhism. I call it the 'hell' of caste because I want to emphasize the utter inhumanity of this vast system of graded inequality.

It's a peaceful revolution that the Triratna Buddhist Community has been supporting for decades. Urgyen Sangharakshita advised Ambedkar in the 1950s on how to convert to Buddhism in such a way that the rest of the Buddhist world would take note. Following Ambedkar's untimely death, six weeks after his conversion, Sangharakshita delivered rousing lectures to tens of thousands of his grieving followers. Nowadays more than a third of the Triratna Buddhist Order is made up of Indians, the vast majority of whom are also followers of Dr Ambedkar.

I have also played a part in this Dharma revolution. In the 2000s I led Buddhist retreats for 'locals' in Bodhgaya, where the Buddha gained Enlightenment. Bodhgaya is in Bihar, many hours by train from Nagpur and other areas in Maharashtra where Ambedkar was active. These locals were from communities of people who regarded themselves as Ambedkar's followers, but had still not been exposed to a living Buddhism.

I remember watching the retreatants arrive. From their weather-beaten faces, I guessed that they had just come from the fields. I remember creating the space to hear their deeply held frustrations, their deeply painful memories in relation to casteism. It took me days of listening to support them to find the life in their anger, which turned out to be a profound sense of dignity, self-respect, and the opportunity to fulfil their potential as human beings. Precious memories.

Ambedkar was born in the late 1800s into an 'untouchable' family and suffered the abuses and oppressions of being an outcaste. Remarkably, thanks to a British Empire grant, he got an education, first in India, then in the USA, the UK, and Germany. When he returned home it was as one of the most highly educated people in India. Instead of establishing a career as a highly paid lawyer, he threw himself into social and political activism for the abolition of the legal, social, and religious basis of 'untouchability'.

His activities included setting up associations and political parties to further the interests of the Dalits, or 'Oppressed'. One particular focus of this was the attempt to gain entry to caste Hindu temples. At a conference in 1935, however, Dr Ambedkar advised the Dalits to abandon their protests. Instead he advised them to leave Hinduism entirely and embrace another religion. For himself, he vowed, 'Though I was born a Hindu, I solemnly assure you that I will not die as a Hindu.'[10]

After independence he was co-opted by Nehru and the Congress Party to shape the new Indian constitution, based on justice, liberty, equality, and fraternity. This was a thoroughly modern constitution, enshrining these values in a non-sectarian way, independent of any particular religion.

In Western ears liberty, equality, and fraternity hark back to the French Revolution. However, Ambedkar stated that he derived them from the teachings of 'my Master, the Buddha'.[11]

Ambedkar saw the Buddha's teaching as a means of individual *and* social transformation: self-development *and* social uplift. He understood the connection between a free and just society and an ethical culture that is widespread throughout society. The law, the courts, and the police can only control the criminal minority, not the whole of society. When the whole of society is corrupt, then the courts and the police will be corrupt too. If there is to be a free and just society, a true 'social democracy', it needs to be founded on ethical principles. The Buddha's teachings on *metta* or loving kindness provide such a basis for social relations. The world can't be reformed except by the reform of the minds of its people.

As Buddhists we need to recognize the power of Dr Ambedkar's example and teaching, which is relevant for the whole world. We need to focus our social and political engagement on this kind of fundamental change in outlook and attitudes, rather than specific outcomes. We need to focus on what underlies the climate and ecological emergency, which is the way people *see* life and *live* life. Only this kind of fundamental shift is congruent with the scale of the emergency. We have a distinctive contribution to make by spreading a different perspective on life. A more ethical outlook will lead to a more just, wholesome, and fulfilling society. The Dharma *can* be a transformative force in society.

After his contact with Dr Ambedkar and his followers, Urgyen Sangharakshita became much more aware of the social dimension of Buddhism. Returning to the West, he spoke of transforming self *and* world. This is what the Buddhist life is all about. Many people are tired of their old self and tired of the world. He offered a kind of koan (a spiritual paradox), that transformation of one isn't possible without transformation of the other.[12]

We tend to think in binary terms. It's either *change the system* or *change must come from the bottom up*. Of course, these aren't mutually exclusive. The development of individuals is fundamental in transforming the world. At the same time external conditions can clearly help or hinder our development. And even with helpful conditions we will still need to exert ourselves. So we need both societal and individual change.

Research supports the effectiveness of nonviolent or civil resistance.[13] There is good evidence that when unarmed civilians use tactics like protests, boycotts, demonstrations, and other forms of mass non-cooperation, they are effective in bringing about social change. These kinds of movements work because they are much more inclusive and therefore tend to be larger. They are visible, which means that they attract more ambivalent members of the population. Once these people join in, the movement automatically has links to the security forces, judiciary, lawmakers, and religious authorities.

In 2000 the Serbian people successfully toppled Slobodan Milošović using nonviolent means. Hundreds of thousands descended on Belgrade to demand that he leave office. Police officers refused orders to shoot into the crowd. When asked why, one police officer said, 'I knew my kids would be in the crowd.'

Nonviolent campaigns are disruptive in ways that are hard for governments to suppress. What's more, if they are successful they are less likely to lead to violence and tyranny. The data are clear: when people rely on civil resistance, their numbers grow. When large numbers of people remove their cooperation from an oppressive system, they are most likely to succeed.

US researchers Erica Chenoweth and Maria Stephan spent two years collecting global data on major nonviolent and violent campaigns for the overthrow of a government or territorial liberation since 1900. To their astonishment they found that nonviolent campaigns were more successful than violent campaigns, 51 per cent as against 23 per cent. Nonviolent campaigns were also more likely to be partially successful, 32 per cent to 10 per cent. And here is the real clincher: nonviolent campaigns were much less likely to fail than violent ones, 20 per cent against 60 per cent.

The trend for nonviolent campaigns to be successful has been increasing since the 1950s. No nonviolent campaign has failed after it achieved the active and sustained participation of 3.5 per cent of the population.

So where do you start nonviolent social change? In the words of Arthur Ashe, the legendary Black American tennis player, 'Start where you are, use what you have, do what you can.' I would add, 'Work with others who are similarly committed.' In fact, *be* the change that you wish to see in the world, which is the subject of the next chapter.

In preparation I invite you to join me in imagining the social conditions that would assist in the growth and development of individuals, in ways that don't jeopardize the safety of this and future generations. What would such a 'Pureland' be like?

Guided reflection: imagining the Pureland
Go to https://www.windhorsepublications.com/audio-resources/audio-resources-for-tbh/

- Purpose of this activity: to imagine a modern Pureland, the social conditions that would assist in the growth and development of individuals in ways that don't jeopardize the safety of this and future generations.
- Time: 15 minutes.

13

Be the change

> We but mirror the world. All the tendencies present in the outer world are to be found in the world of our body. If we could change ourselves, the tendencies in the world would also change. As [a person] changes [their] own nature, so does the attitude of the world change towards [them]. This is the divine mystery supreme. A wonderful thing it is and the source of our happiness. We need not wait to see what others do.
>
> – M.K. Gandhi[1]

You might be familiar with the saying, 'Be the change you wish to see in the world.' This is usually attributed to Gandhi, though an online search reveals that the credit actually goes to Arleen Lorrance, a high school teacher from New York. She wrote, 'One way to start a preventative program is to be the change you want to see happen.'[2] By degrees, this became the 'Be the change' quote that we're familiar with, and was misattributed to Gandhi, who did indeed say something similar.

In the previous chapter we heard the story of the babies in the stream that I got from Marshall Rosenberg. Marshall, as I call him, was very interested in the spiritual basis of social change. Echoing Gandhi, he said that he tried to change the paradigm, the social structure, *within himself*. He was concerned that unless as social change agents we come from a spiritual perspective, we're likely to do more harm than good.

Spirituality, for him, was staying connected with our own life and to the lives of others, moment by moment. It was important to start with ourselves – but not end there. This is because spirituality can be 'reactionary' if people become so calm, accepting, and loving that they tolerate dangerous societal structures. The spirituality that we need to develop for social change is one that mobilizes us for social change. It's not about just sitting there and enjoying the world no matter what. It creates a quality of energy that mobilizes us to action. He said that unless our spiritual development has this quality, 'I don't think we can create the kind of social change I would like to see.'[3]

Marshall cut his teeth as an activist in the US Civil Rights era. He was involved in preparing white and black communities in the South for 'desegregation' – coming together in schools and other institutions. He found that when people used labels such as 'black' or 'white', 'good' or 'bad', 'right' or 'wrong', they reinforced their sense of separateness. By contrast, when he brought people's attention to their feelings and needs, it became easier for them to understand each other and recognize their common humanity.

Marshall saw himself as following in the footsteps of Gandhi and Martin Luther King. We'll come to Gandhi in the next chapter. In August 1967, less than a year before his tragic death at the hands of James Earl Ray, King announced, 'I have also decided to stick with love . . . I've seen too much hate on the faces of sheriffs in the South . . . Klansmen . . . Councillors . . . I know that it does something to their faces and their personalities, and I say to myself that hate is too great a burden to bear.' Towards the end of the speech, he declared, 'Let us realize that the arc of the moral universe is long, but it bends toward justice.'[4]

In arguing that hatred will never conquer hatred, King echoed the Buddha. As the Buddha said, only love can do that. In the same speech King affirms that we will reap what we sow. This amounts to a kind of justice, albeit a different kind from the punitive sort that is primarily enshrined in our legal systems. We will return to the topic of climate justice in the next and final chapter.

Marshall named his process *Nonviolent* Communication to recall Gandhi's and King's nonviolent social change programmes. Like most Westerners, Marshall wasn't aware of Dr Ambedkar's contribution to nonviolent social change. It was only in the early 2000s, when he came in contact with members of the Triratna Buddhist Order in the West, who were then in India, that he even heard his name. However, once Marshall heard of the Dalits' struggle, he enthusiastically supported it, and in 2006 led a retreat for 3,500 participants.

Marshall said that if we want to change society, we need to be the change that we wish to see in the world. Social systems mirror social relations, and social relations mirror the kinds of thinking that exist in the minds of individual citizens. So we need to change the 'social structure' within ourselves to something that is more life-serving.

Marshall saw in himself and his activist colleagues that social change efforts tend to focus on what Joanna Macy called 'holding actions'. This is when we try to stop people from doing things that are experienced as harmful by others. With holding actions, there's a risk that we end up trying to fight the existing reality. As we saw in the previous chapter, if you want to change something, build a new model that makes the existing model obsolete. In terms of the parable of the burning house, this new model is expressed in the cry of inspiration that persuades the children to run to safety. Marshall's cry of inspiration was to *be* the change.

For the record I do believe that the climate and ecological emergency is an ethical issue, and I want to respond in an ethical manner. However, I'm also aware that this approach has a limited appeal, and has been tried already. So we need a cry of inspiration, to *be* the change and show a different way that is less costly and more fun. In my experience, this is usually more of an effort than standing on the street and holding up a sign saying 'Stop!' Despite this, transforming the social structure within ourselves is more likely to result in sustainable social change.

The process takes us all the way from self-development to social change, while removing blocks to action along the way. Since social change is a collective process, you might want to work through the rest of this chapter with a small group of friends or colleagues or think about what you could do with friends and colleagues after reading it. Most people find it helpful to write things down as they go along, to track the process.

The process begins with identifying a sub-world that you want to affect, explores your strongest judgements, creates the possibility of 'owning' these judgements, then connects these judgements to their underlying feelings and needs. In the middle part of the process, you imagine the social practices that would be in keeping with your needs, and find a way to show these to others. The process ends with concrete next steps to make your vision reality, and preparation in the form of empathizing with the current practices.[5]

I suggest that you read it through first, then follow the guided reflection at the end of the chapter. Whether you're reading or following the guided reflection, I suggest that you refer to the lists of feelings and needs at the end of the book. In order to prepare and ground yourself, you could listen to the 'arriving exercise' from the introduction or 'coming to our senses' from chapter 2. And, of course, if you'd like to go through it with me in person, you can join one of my Burning House courses.[6]

Identify a sub-world. Begin by identifying a social system, a world, or a sub-world that you want to affect today. To keep your attention on learning the steps of the process, I recommend that you choose a social system, world, or sub-world where you have some power. This is your opportunity if you want to bring your environmental concerns into your family or partnership, campaign group, a work team or company, class or caste, neighbourhood, or institution.

Identify your strongest judgements. In order to transform our inner social structure, we need to become aware of our judgements about this sub-world in relation to the climate and ecological emergency. We need to access and transform the vast amount of energy that is trapped in the lattices of these static judgements. This step parallels the work we did in chapter 10, where we explored ecological grief. Welcome your thoughts about this world or sub-world. Write them down. Here are mine: *We're destroying the planet. No one seems to care. It's just business as usual here. I might as well not bother.*

Now, what are you feeling most frustrated, angry, guilty, and depressed about? *My own utter powerlessness. From the seventies onwards, oil companies did research on the impact of fossil fuels on the atmosphere. They knew what they were doing, but they went ahead and did it anyway. It was a callous disregard for the whole of life on the planet. And we're still doing it. And I'm still part of it! Arrgghhhh!!!*

To connect with your thoughts energetically, I suggest that you say each thought out loud with the energy you feel. *Arrgghhhh!!!*

Owning your thoughts. The purpose of this step is to take full responsibility for the story that you're telling yourself. This may come as a surprise, but your thoughts don't have the status of universal or universally held truths! Try repeating each thought with 'I'm telling myself . . .' in front of it. *I'm telling myself that I'm utterly powerless. I'm telling myself that from the seventies oil companies did research on the impact of fossil fuels on the atmosphere. I'm telling myself that they knew what they were doing, but they went ahead and did it anyway. I'm telling myself that it was a callous disregard for the whole of life on the planet. I'm telling myself that we're still doing it. I'm telling myself that I'm still part of it.*

Now you. Repeat the sentences that have the most charge. *I'm telling myself that I'm utterly powerless. I'm telling myself that it was a callous disregard for the whole of life on the planet.* Keep repeating them. For some people, this kind of mindfulness or awareness exercise really works. Recognizing thoughts *as thoughts* in this way can bring an unexpected, significant shift in their experience. It supports acknowledging that it's a *story* that we are telling ourselves, a partial rather than a whole truth.

Identifying your feelings. Connect each thought with a feeling, as we did in chapter 11. To do this, ask yourself, 'How do I feel when I tell myself such and such?' Notice your breath and other bodily sensations. Write down, 'I am feeling . . .'. If you're struggling and need a prompt, you could read the list of feelings at the end of the book and sense which one(s) resonate in your body. *When I tell myself that 'I'm utterly powerless' I feel heavy, and at the same time puzzled and hopeful. When I tell myself that 'it was a callous disregard for the whole of life on the planet', I feel deeply sad and tearful.*

Connecting with the underlying needs and values. The purpose of this step is to ground your activism in your deepest values. Everything we do is in the service of our needs and values. Earlier we encountered Max-Neef's list of fundamental human needs: subsistence, protection, affection, understanding, participation, leisure, creation, identity, and freedom. Max-Neef added the possibility that a tenth, transcendence, may in time become a universal need.[7] I suggest that you check in your body and see what resonates with you. If you need support, you could make use of the list at the end of this book.

We are always trying to enrich our lives, even when those strategies fail. Feelings are feedback about our needs and deeper motivations, so connect each significantly different feeling with an underlying need. Write down, 'I am feeling . . . because I need/value . . .'. *I am feeling heavy, because I need and value a sense of empowerment, a sense of choice, agency, and connection to life. By contrast, I am feeling puzzled because I want to understand how I became so fully convinced of this story. With this, I'm feeling hopeful because I value empowerment, choice, and agency. At the same time I am feeling deeply sad and tearful because I love life, I love the earth, and I want this and future generations to have the chance to live in health and safety.*

If you're working in topic groups, I suggest that you share your process so far, focusing especially on the needs and values that you're now in touch with.

Envisioning. Now the magic part! The purpose of this step is to vividly imagine the social practices in keeping with the needs and values that you have just identified. This step corresponds to the father's cry of inspiration in the parable of the burning house. We can imagine him describing in vivid detail the toys that are waiting just outside. So how do we create this vivid picture of a more beautiful world?

You could try this exercise: imagine for a moment that you've been asleep and dreamed of a more beautiful world. When you wake up in the morning, something has changed. You look around for clues as to what it is. With a shock of delight, you realize that you are already in the world that you want to live in. Now look around: what are the people saying to each other in their interactions? What are they doing? What are they putting their energy into? What are they using their resources for? What are the new social practices? I suggest that you make a note of these.

Try to find a way to *show* the new world that you dream of. This could be a mime, a monologue, a short play with words, a song, a poem, a statue, or a combination of any of these.

Take the time you need to represent the different aspects of your dream.

Showtime! The purpose of this step is to bring the new world alive by showing it to others, giving them a sense of what this world looks, sounds, and feels like. To create a sense of atmosphere and occasion, you could designate a stage area and put up a sign saying Showtime! If at some point you come to do this with a group, you can invite topic groups to come up one by one and show their dream. To respect the flow of creativity, encourage people to give feedback at the end of all the presentations, rather than after each one. The feedback questions could be: what touched you about this dream? What did you learn?

Memories flood in of the many times I've witnessed Showtime! with groups in the UK, India, Sri Lanka, Nepal, Australia, Germany, and the USA. How people show their dreams varies according to their culture. Some of the most memorable dreams have been communicated without words. One group in India mimed a young woman in distress, wanting to marry someone outside her caste, being supported by her family. The group swirled in support around the young woman, eventually inviting the audience to rise up and join in. It was a kind of collective epiphany.

During the civil war in Sri Lanka I led a mixed group of Sinhalese, Tamils, and Muslims through this process. Most of them were in their early twenties. When we came to Showtime! they presented a short play, imagining themselves fifteen years older, married with children. They talked of arranging play dates for their children, and even considered intermarriage between them. One of my co-trainers was genuinely shocked to witness this. 'What did you do with them?' she asked. 'I only took a break for a couple of hours, and when I come back, they're talking about letting their children intermarry?' She couldn't imagine what had changed to bring about this degree of reconciliation. I put it down to connecting with the underlying needs and values, then the magic and power of working with dreams in Envisioning.

Next steps. The purpose of this step is to create a bridge from the current world to the new world of the dream. It also helps to ground the process in concrete realities. What requests or offers will make your vision reality? Create an action plan in your topic group. If you make requests of yourself or others, can you make them SMART (specific, measurable, achievable, realistic, time-bound)? What's the first step towards making your vision reality? If that seems too big a step to take, identify a smaller step towards it. As the *Tao Te Ching* says, a journey of a thousand miles begins with a single step.

For instance, this step has led me, as a Buddhist mediator and activist, to learn and experiment with new paradigms in community building. I've learned that the new paradigms need to cover the overall purpose of the group, support systems (including healing), learning and information transfer, decision-making, conflict transformation, fun/play/rest, and resource allocation or economy. This has led to designing a Year Training with modules addressing each of these topics. I'll describe all seven briefly.

Every group, community, or organization needs a *purpose,* something that draws it together. To make this clear, can it be expressed in plain language in one sentence? To honour interconnectedness, is it easy for anyone to see how this purpose contributes to life? Groups that have such a clear purpose are more likely to survive and thrive.

Secondly, groups need some form of *support network* of peer support. In a Buddhist context we have chapters or *kulas* (groups) that meet regularly to practise together. In an NVC context we have empathy groups and practice groups. In Extinction Rebellion there are the different affiliation groups. These groups can offer each other support, and sometimes challenge! This can include the more specialized form of support in healing past pain, which can get in the way of our current plans.

Thirdly, groups or communities need some system for *learning* and information transfer. This can be in the form of peer learning, training from a 'trainer', or self-study. It can be in person or online. If it involves exercises, these can be more structured – *Here's what we're going to learn and here's how we're going to learn it* – or less structured – *Let's do this activity and see what each of us learns from it*. Any learning system will include a process for giving and receiving feedback, both of which are essential for learning.

Fourthly, a group or community needs a way of *making decisions*. As an alternative to the usual 'majority takes all' decision-making system, I have been exploring the principle of *consent*. Consent means that there are no active objections to the proposal on the table. Consent decision-making holds out the promise that everyone's voice is valued, and that we get to make decisions and organize a project together rather than getting bogged down in endless discussion. Consent is a precious commodity in any group, and having a clearly defined and understood decision-making process can support this.

Fifthly, we need a process for *conflict transformation*. We need a process that is essentially *restorative* (restoring trust, connection, and safety) rather than punitive (you've broken the rules and now we're going to punish you for it). This has led me to learn and practise 'restorative circles', a process developed by Dominic Barter and associates in the slums of Brazil. Restorative circles apply the principles of Nonviolent Communication to restoring trust, safety, and connection when these have been breached in communities.

Sixthly, we need space and ideas for fun, play, and rest. On the current NVC Year Training that my partner Gesine and I are supporting, the whole group has appointed a Minister of Fun to remind us of the importance of these!

Seventh and finally, we need a process for *resource allocation*. In other words, we need a different *economic* paradigm to help us move away from making money a measurement of worth and value, towards focusing on the sustainability of the community as a whole. In the Buddhist Sangha we've been practising dana (generosity) for over two thousand years. In the NVC community we've been experimenting with gift culture, an updated version of dana, which can work for businesses as well.

Preparation: empathizing with the current practices. The purpose of this step is to prepare the ground for the requests and offers in the previous step. It's a way of embedding the new practices and preparing for the new world. Practise empathizing with the needs that are being met by the current practices. Imagine asking the person who is currently doing something that you experience as harmful, 'Are you needing . . . ? Do you value . . . ? Is . . . important to you?' Write it down. In a role play, if you can, ask the person, 'Are you doing this because . . . is important to you?'

My mind flashes back to the first time I saw Marshall lead a social change workshop, in London. It must have been 2004. He was coaching a group of climate activists in how to speak to oil company executives. For this step in the process he set up a desk on the stage and took the role of the executive. One by one the activists came up onto the stage, sat down in front of his desk, and tried to engage with him.

As soon as they opened their mouths, out came strong moralistic judgements about oil company executives and their motivations, 'Ignorant bastard, selfish swine, evil bastard.' I watched in awe as each one crashed and burned. Their collective 'enemy image' of the executive was so strong that none of them could connect with him as a human being. None of them was able to engage him in a conversation about what his company was doing to the planet. Marshall had to work very hard, stepping out of his role in order to offer *them* empathy and coach them in empathizing with him!

Reflect on the process. The purpose of this step is to harvest the embodied learning from this process. Where are you now, physically, emotionally, and metaphorically? What have you learned? What are you taking away with you? I suggest that you write it down and share it with your topic group or the group you've been working with. Watching Marshall's role play, I resolved to do the inner work that would enable me to talk to people in positions of power as fellow human beings, with feelings and needs, while staying true to my values.

In the next chapter we're going to look at climate justice and the case for nonviolent disruption as a means of supporting it.

Meanwhile, I invite you to join me in opening your heart and allowing love in!

Guided reflection: be the change!
Go to https://www.windhorsepublications.com/audio-resources/audio-resources-for-tbh/

- Purpose of this activity: to deepen your understanding of what it means to *be* the change, by transforming the 'social structure' within and showing a different way that is less costly and more fun.
- Time: 25 minutes.

14

Climate justice and nonviolent disruption

> If it remains unresponsive to those wishes, or not sufficiently responsive – and the situation is one of extreme urgency, where every day is precious – then more serious measures should be taken and pressure brought to bear on the government by means of mass civil disobedience along Gandhian lines.
>
> – Urgyen Sangharakshita[1]

In 'Buddhism, world peace and nuclear war' Urgyen Sangharakshita suggests various courses of action in order to eradicate nuclear weapons and thereby the possibility of nuclear war. In parliamentary democracies they include the obvious, such as voting, lobbying members of parliament, petitions, meetings, marches, demonstrations, and so on. If these fail, and the situation is one of extreme urgency, he suggests more serious measures such as mass civil disobedience along Gandhian lines.

What could this mean? Fortunately we have the transcript of a seminar in which Sangharakshita describes Gandhi's Salt

March, the real beginning of Gandhi's influence in India.[2] Gandhi was part of the Indian self-rule movement, which wanted to remove the British colonial government. We know that he chose the issue on which to campaign nonviolently very carefully.

The British administration had placed a tax on salt. They also regulated the manufacture of salt by permitting it to be manufactured only under licence. Controlling both ends of the process gave the British government an effective monopoly on this essential item.

Gandhi reflected that salt was something very basic, which even the poor had to use. At the time most Indians were villagers, working in the fields. When they worked hard they sweated and lost a lot of salt, which they needed to replace. So they couldn't possibly do without it. When they bought salt, they paid tax to a *foreign* government. This was an issue that affected everybody and concerned everybody, so Gandhi could rely on mass support.

Gandhi led a march of thousands of people from his ashram in Ahmedabad, Gujerat to the seaside, where they began manufacturing salt without a government licence. The leaders got arrested and hundreds of nonviolent protestors were beaten by British police. The march received worldwide attention through extensive newspaper and newsreel coverage.

Gandhi was successful in challenging British rule because he dramatized the issue that 'this foreign government is taxing even the salt on our food, salt that even the poorest need, and need more than anybody, because they are doing hard physical labour'. The Salt March is an example of successful symbolic nonviolent civil resistance. It helped to swing the tide of public opinion in India, where millions copied his actions, then in the UK and the USA, where it swung public opinion against British rule and in favour of independence.

If we are to follow Sangharakshita's suggestion, before considering mass civil disobedience along these lines we need to establish the case for climate justice *and* the extreme urgency of the situation. I hope that the rest of the book has established the case in intellectual, emotional, and social terms.

Considering climate justice, we can take heart from a couple of Buddhist scriptures that deal with this aspect of government. Both argue that a primary cause of theft and violence is poverty. The *Kutadanta Sutta* describes a kingdom in which theft and crime have increased to the point that people are afraid to walk the streets for fear of violence. The king considers increasing fines and punishments for the wrongdoers. He also plans a great animal sacrifice and goes to a holy man for advice. However, he doesn't get what he's expecting. The Buddha-like holy man argues that increases in punishment will be self-defeating. It will only increase the sickness of society because the root causes will remain untouched. The root causes are economic injustice and poverty. He says that to stop crime the king needs to improve economic conditions by giving food and seed corn to farmers, capital to traders, and food to those in government service.[3]

Another sutta, the *Cakkavatti Sihanada Sutta*, makes a similar point. The duties of a king are to prevent wrongdoing in the realm and to give wealth to the poor. Those kings who keep to this are blessed with peace. However, one king who neglects the poor finds that his realm descends into chaos. Poverty becomes rampant and this leads to theft, as people would rather steal than die. This eventually leads to a week of slaughter in which the citizens kill each other with swords, like wild beasts.[4]

Inevitably the impacts of the climate and ecological emergency will be more serious for those humans and animals with fewer resources, less of a cushion to adversity. For instance, the oceans, the poles, and equatorial regions are experiencing the brunt of global temperature increases. And the 2018 Intergovernmental Panel on Climate Change (IPCC) *Fourth National Climate Assessment* report found that low-income individuals and communities are more exposed to environmental hazards and pollution and have a harder time recovering from the impacts of climate change.[5] Care for all beings also means a concern for climate justice, a concern for these disproportionate impacts.

Take India, for example, home to the Buddha. The World Bank's predictions, based on an average global temperature rise of 2–4°C, are far more frequent heatwaves, changing monsoon patterns, droughts, glacial melts, water shortages, and sea level rise, all affecting agriculture and the health of the population and leading to migration and conflict.[6] We are back in the realm of the *Kutadanta Sutta* and the *Cakkavatti Sihanada Sutta*, without a king or indeed any institution, even the World Bank, that can take the kind of actions that would address the scale of the emergency.

Historically, Western industrialized nations have been responsible for the great majority of the carbon in the atmosphere, though they are fast being caught up by China, and increasingly India. By contrast, poorer nations in equatorial regions, many of them stripped of resources and dignity by centuries of Western colonialism, are going to experience the greatest impacts. The implication is that the nations that will suffer the greatest impacts have contributed to and benefited least from the carbon we have dumped in the atmosphere.

So we have disproportionate impacts, and disproportionate contributions and benefits. These call for equity, for climate justice. And then there is intergenerational climate justice. Older generations have contributed to and benefited more from the causes of the climate and ecological emergency, while they have a shorter time to experience the consequences. For younger generations it's the opposite. As Greta Thunberg told heads of state gathered in New York for the 2019 UN climate action summit, 'This is all wrong. I shouldn't be up here. I should be back in school on the other side of the ocean. Yet you all come to us young people for hope. How dare you. You have stolen my dreams and my childhood with your empty words. . . . We are in the beginning of a mass extinction and all you can talk about is money and fairy tales of eternal economic growth – how dare you!'[7]

In a Buddhist context, the notion of climate justice inevitably comes back to a discussion of karma. Karma is a non-punitive form of justice. There is no transcendental judge in the sky who is measuring our actions and handing out appropriate consequences. Instead, actions have consequences according to the volitions with which they are performed. Actions performed out of greed, hatred, and ignorance inevitably lead to further suffering, for oneself and others.

We can claim that we just weren't aware of the likely impact of our actions. Certainly, anthropogenic global warming wasn't common knowledge until the third IPCC report in 2001.[8] Prior to that, it appears that oil company executives deliberately concealed research that would have raised the alarm. At the same time, that doesn't mean that we will escape the consequences. The earth's biosphere is nothing if not a massive feedback loop. It envelops the earth and gives us feedback on our collective actions.

If we are going to talk of climate justice, we need to be clear that this is a non-punitive form of justice, rather like the restorative justice that I touched on in the previous chapter. Restorative justice is concerned with restoring, and even increasing, the safety and wellbeing of the earth and its citizens. I support those who are bringing climate lawsuits and campaigning to make ecocide (the mass damage and destruction of ecosystems)[9] an international crime, as these are important to establish the notion of climate justice in different legal contexts.

One way of approaching climate justice is to talk about rights – to clean air, water, etc. There are problems with this approach. Firstly, rights imply corresponding duties to uphold those rights. So who has these corresponding duties? We would need to identify a person or body to do this, otherwise the notion of a right is meaningless. Secondly, rights tend towards a punitive form of justice, as they seem to require enforcement.

In order to measure wellbeing, I'd like to suggest a needs-based approach, as outlined in previous chapters. Humans and most other sentient beings need subsistence, protection, affection, understanding, participation, leisure, creation, identity, and freedom, with a possibility that a tenth, transcendence, may in time become a universal need.[10] I understand that this is just a suggestion, and it needs to be fleshed out in philosophical and legal contexts.

In any case, addressing the climate and ecological emergency requires something more difficult than punishing the 'offenders'. As we saw towards the end of the last chapter, it involves empathizing with the fears and unmet needs that have provided and continue to provide the impetus for this emergency.

So there is a case for climate justice, though we need to define this carefully.

Sangharakshita's second criterion is that of extreme urgency. Is the climate and ecological situation really an emergency that justifies civil disobedience along Gandhian lines? And have all other avenues already been exhausted? In September 2018 the UN Secretary-General António Guterres called on 'friends of Planet Earth . . . to sound the alarm' and described the climate and ecological emergency as 'the defining issue of our time' since we face a 'direct existential threat'. He continued, 'If we do not change course by 2020, we risk missing the point where we can avoid runaway climate change, with disastrous consequences for people and all the natural systems that sustain us.'[11]

2020 has come and gone, and our course has changed very little. CO_2 concentrations are accelerating, we are adding heat to the oceans at a massive rate, the polar ice caps are melting. The Great Barrier Reef has suffered three major bleachings in the past five years. Fires set by humans are destroying the remaining Amazon rainforest. Wildfires are destroying the Australian bush and Californian forest.

In an emergency we need a full-on emergency response. However, governments' promises of reductions are still mainly promises rather than regulations and funded policies. Reductions in emissions due to the Coronavirus pandemic have only been temporary and unsustainable. Before that, conventional campaigning in the last thirty years has produced very little effect on government policies and business practices. In other words, not enough is being done to avert climate catastrophe.

As veteran climate campaigner Jonathan Porritt, former director of Friends of the Earth, puts it, in order to avoid runaway climate change 'there is no hope whatsoever in another ten years of incremental change'. He advocates mass civil disobedience, 'to force our politicians to step up and do what is now needed'.[12] We need to acknowledge the urgency of the situation and act on it. Otherwise we are complicit in it. Sometimes we need to use force in a protective manner in order to protect needs. Not to punish or to teach a lesson, but in the service of compassion.

David Loy, Buddhist teacher and author of *Ecodharma*,[13] points out that in the modern world, greed, hatred, and delusion have become institutionalized. Corporate consumer capitalism is institutionalized greed. Militarism is an example of collective, institutionalized hatred, and the media, with its fake news, is institutionalized delusion. We need to work together with others to counteract these forces, which have taken on a life of their own. He argues that nonviolent disruption has a place in this, alongside systemic change and transforming consciousness. 'Nonviolent disruption shakes things up and draws attention,' he told me. 'Modern life can be pretty comfortable, especially if you're white, male, and middle class like me. We need actions that will disrupt business as usual.'[14]

In other words, we need more serious measures. Of course it's important to remember that nonviolent disruption isn't the only avenue that remains open to us. Other mass protests such as the school strikes have been remarkably effective and are likely to remain so.

What kind of actions are we talking about? Remember that Gandhi thought long and hard before finding a dramatic and symbolic issue on which to challenge the British administration. When the Buddha heard that his relatives were going to war with each other, he simply put himself between battle lines and started asking both sides what the dispute was about. In one of the great ironies of the Buddhist scriptures, they couldn't tell him! The Buddha also sanctioned monks and nuns to release animals from hunters' traps, if they were acting out of compassion.[15]

In 2019 there was a sea change in public opinion about the climate and ecological emergency. This was represented by increased media coverage, a change in *how* the media reported it, and people re-evaluating their climate-impacting choices both individually and collectively. Greta Thunberg's school strikes and Extinction Rebellion (XR) can take some of the credit for this sea change. They were out there, grabbing the headlines and people's imaginations. I was attracted to XR's explicit commitment to nonviolence, including an attitude of 'No blame, no shame' towards those responsible. We are all responsible. We have all benefited from the ease and comfort of our fossil fuel-driven economies.

With slogans like 'Tell the truth' and 'Act now' Extinction Rebellion used language that was congruent with my own sense of urgency. I didn't enjoy how they phrased these as 'demands' – my understanding is that demands are an expensive currency. And the idea of a rebellion stuck in my throat. I followed Marshall Rosenberg in this, who advised us never to give other people the power to make you submit *or* rebel. In other words, always keep a sense of your own power and autonomy.

However, as soon as I offered XR my skills in nonviolence and de-escalation training, I felt a deep sense of relief. Doing something counteracted my habitual sense of helplessness and inertia. There were many moments when I felt deeply connected to my humanity and my compassionate intention.

One such moment was outside Downing Street in October 2019. I'd finished my shift outside the Home Office and was looking around some of the other XR sites. As I walked along Whitehall I came across a group of fellow de-escalators and stood with them in a silent circle for a couple of minutes. Further up, outside Downing Street, there was a samba band going at full pelt, hundreds of 'Rebels' and a line of policemen preventing people from coming away from Trafalgar Square. Within this chaos I spotted some XR Buddhists meditating. I sat down and watched them for a while, enjoying their peaceful expressions amid the din.

While I was sitting there I noticed a huddle of policemen behind me. Feeling curious, I went over. Between the policemen was a row of four people lying down. Someone told me that they were part of a die-in that had happened earlier in the day. Memories came back of my first experience of mass civil disobedience in the mid-1980s, outside the Oxford Union.

There was a kind of symmetry. Back then, Boris Johnson had been inside the Union building, waiting for the junior minister responsible for nuclear power to arrive. Now, presumably, he was sitting in Number 10, waiting for the samba band to quieten down! And I was still outside with the Rebels, trying to disrupt business as usual. Some things never change!

The heavens had opened since their die-in, and the protestors on the ground looked wet and bedraggled. I asked them if they could do with a warming drink. Some said yes, a black coffee, and a coffee with soya milk and sugar. At this point I'd hardly looked at the policemen. However, one of them caught my eye and said, 'That's eight coffees.' He smiled. I looked around and noticed that the other officers were smiling too. I grinned back, recognizing them as people who had been standing for hours, and who would possibly continue standing for hours, until they were relieved.

As I walked to Westminster Underground station for the coffees I reflected on how easy it is to identify with one group, one gang, and see the other group as the enemy. Now I don't want to whitewash the police, in the UK or elsewhere. I've seen and heard too many credible accounts of them using their individual and collective power in ways that don't respect the safety and dignity of the citizens they are supposed to be responsible for protecting. However, I resolved to try to see the police as much as possible as people doing their job, with families to support, ideally as protectors of public safety. And to see myself as a protector too, a conscientious protector of the planet and the lives of future generations.

A couple of days later I spotted Yogaratna, a fellow Triratna Order member, in a line of meditators being arrested.

I was curious to understand more about his motivation, so afterwards I caught up with him online. I reminded him of the moment I'd seen him meditating in the crowd. He remembered that it had been really ranty, and then the mood suddenly changed to a peaceful chanting of 'Om mani padme hum', a mantra evoking universal compassion.

I said, 'That was me. I started the mantra.' I'd seen the line of meditators and noticed the noise around them was deafening, so I approached the man with the megaphone and told him that I had something more peaceful to offer. When I started chanting 'Om mani padme hum', the people around me joined in.

Yogaratna looked astonished. 'That's really magical, to hear that,' he said. 'When the mood suddenly changed it touched me very deeply.'

I asked him about his motivation. He said, 'I think that things are so urgent today that people of faith in particular need to speak out. It's important that we do this from as resourced and skilful a place as possible, so we need to meditate and go on retreat and so on.'

And what attracted him to nonviolent disruption? He said, 'Nonviolent civil disobedience was central to the success of many other social change movements – the suffragettes, the civil rights movement in the USA, even Indian independence – that we have come to regard as justified. At the time, in some quarters these movements were extremely unpopular.'

Protest can really make a difference, he continued. Does anyone seriously think that women would have been 'given' the vote by the men who had the power to grant it to them, unless women had spoken out, argued, and campaigned for it? When people of all kinds – grandmothers, vicars, doctors, young people, ex-police officers, shop workers, nurses – are willing to risk arrest, it does inspire respect. Protest can irritate people, but it can also get genuine discussion and dialogue going.

If, like Yogaratna, you are planning to engage in nonviolent disruption, I suggest bearing in mind three principles: accountability, vulnerability, and acting from community.[16] Accountability means staying and explaining why you have done something. Running away short-circuits whatever power you may have in the situation. Accountability means being willing to accept the consequences of your actions. Legally it means being willing to take the rap for your actions, and ethically it provides restraint and shows that you yourself are willing to experience disruption.

Vulnerability stems from accountability. In staying and explaining, you become present with your vulnerability. This can be extraordinarily powerful. If you are really willing to make yourself vulnerable for what you believe in, this becomes a measure for others of your sincerity and integrity and the importance of the issue. You don't need to have all the answers. You can see action as a dialogue and be willing to search more deeply into your own complicity. We've *all* benefited from what has led to the climate and ecological emergency.

Acting from community provides a democratic check on nonviolent disruptive actions. Are you really acting *from* community? Community will also be a source of support as you go into risky situations. We all have responsibilities as citizens to each other. The climate and ecological emergency is too important to leave to politicians, who have other concerns, like getting and staying elected. Participatory democracy means showing up as concerned citizens, and it strengthens our representative democracy.

Acting from community is familiar to members of the Triratna Buddhist Order. In 1993 Urgyen Sangharakshita encouraged Order members to 'remember that you are a citizen'. Followers of the Buddha are also citizens of a particular country. As citizens, alongside inevitable disadvantages, we enjoy certain benefits, which vary from country to country. Sangharakshita encouraged Order members to feel grateful for these benefits and to make a contribution to that society. For instance, centres can have a positive influence on local life, and Order members can get involved in local government and make a positive contribution. He concluded, '[R]emember you are a citizen. You have a responsibility towards your fellow citizens as well as a more spiritual responsibility to the Movement and your fellow Order Members.'[17]

So what are the limits to nonviolent disruption; when does it become unskilful? To predict the consequences of an action, we need to look at its motivation along with an awareness of the likely consequences. We can do this from our imagination and our experience.

Awareness has many aspects. If you're not careful, breaking the law could lead to undermining the rule of law, which is generally protective. Undermining the rule of law could lead to a breakdown in the fabric of society. In any case, disruption can have a tendency to slide towards violence. From my own experience, things can escalate very quickly in a crowd, hence the importance of nonviolence and de-escalation skills to maintain 'nonviolent discipline'.

One XR action that went wrong created a lot of adverse publicity. Rebels climbed on top of Underground carriages at Canning Town and unfurled banners. It was a weekday morning, and the relatively low-income inhabitants of the area were trying to get to work. There were scuffles between protestors and passengers on the platform, who told the protestors to get down and let them get to work. The video of this incident was watched more than a million times on YouTube and news of the incident may have led to a reduction in support for XR among the general public.

So what happened? Firstly, the Rebels targeted public transport, which is already sending out a mixed message. Any version of a sustainable or even survivable future involves using public transport more rather than less. Secondly, their action impacted working people, rather than politicians and corporations, who have far more power to effect change. Thirdly, the action wasn't safe: the protestor could have been seriously hurt when he was pulled from the roof of the carriage and beaten. Fourthly, most XR members were highly critical of the action when it was proposed, so the initiators didn't act 'from community'.

Can a sense of trust and safety be restored after such an incident? My experience of hosting restorative dialogues tells me that it can, with a willingness to listen on both sides, and that those involved are willing to acknowledge the full impact of their actions and demonstrate with concrete actions that they have learned from their experience.

When you are planning nonviolent disruption it's important to find the nonviolent sweet spot. How do you get people to realize the magnitude of the dangers we're facing? Psychologically we're wired to turn away from dangers that are too great to countenance. Harking back to the parable of the burning house, you will need a cry of anguish followed by a cry of inspiration. If you speak too quietly, no one will be able to hear you. If you break something precious and create revulsion in the onlookers, they will probably shut down and be unwilling to listen to the underlying message.

Finding the nonviolent sweet spot means doing something that disrupts business as usual enough to create space for your alternative message. People are unlikely to consider the vision unless they are also aware of the danger. Part of the cry of inspiration could be offering more reflective spaces, perhaps meditation, to cultivate more skilful motivations and to support people in reflecting on their experience.

Finally, a health warning: deliberately breaking the law involves risking your liberty and wellbeing and is likely to impact those around you. If you are thinking about risking arrest, I recommend that you get informed about the likely legal consequences and consider how it might impact your life more broadly. Remember that there are many other ways to support nonviolent direct action. I chose to offer my nonviolence and de-escalation skills, based on Nonviolent Communication, and to use my writing skills here.

Considering climate justice and nonviolent disruption can get rather heady and take us away from our hearts. As we come to the close of the book, I invite you to join me in opening our hearts to the world. Nothing forced, just sensing love where it's there for us, and allowing it in and through.

Guided reflection: allowing love in – and through!
Go to https://www.windhorsepublications.com/audio-resources/audio-resources-for-tbh/

- Purpose of this activity: to open your heart to the world by allowing love in and through.
- Tools: somewhere to sit or lie down quietly and comfortably, a notepad or journal, and a pencil (optional).

15

Final thoughts: the beauty and the terror

> If we continue abusing the earth this way, there is no doubt that our civilization will be destroyed. This turnaround takes enlightenment, awakening. The Buddha attained individual awakening. Now we need a collective enlightenment to stop this course of destruction. Civilization is going to end if we continue to drown in the competition for power, fame, sex, and profit.
>
> – Thich Nhat Hanh[1]

It can be helpful to go back and look at how we dealt with previous emergencies as we turn to face this one. In the 1980s Urgyen Sangharakshita wrote a booklet entitled 'Buddhism, world peace and nuclear war'.[2] Much of what he says about the threat of nuclear war can be applied directly to the current climate and ecological emergency. Both are *global* threats, and we can only address them if we realize that humanity and the natural world are indivisible and consistently act on that realization. We belong to the whole of the earth, rather than any particular section of it.

We are responsible for it all. This is the voice of sanity and compassion. This is what is reasonable and valuable. By reminding ourselves in this way, we may even be able to touch on what is most valuable in our lives.

When we bring pressure to bear on our governments, I suggest that we make it massive, unanimous, unmistakable, and strictly nonviolent, until the threats are averted. Apart from bringing pressure to bear on governments, we also can bring pressure to bear on our fellow citizens, especially those in our own national community. We can do this by disseminating information about the extent of the threats and helping people to develop a more positive attitude towards other national communities, species, and ecosystems.

We can encourage our fellow citizens to learn more about the lives of other nationalities and species. We can teach our fellow citizens the *metta bhavana* (loving kindness meditation). We can model sanity, compassion, and a love of humanity and of the earth as a whole. In order to do this, I suggest that we intensify our commitment to the great ethical and spiritual principle of nonviolence or love.

The climate and ecological emergency, like the threat of nuclear war before it, has shown us the consequences of collective violence (greed, hatred, and delusion) towards the earth, and given us a much deeper appreciation of the real value of nonviolence. Can we imagine a world in which our relation to the earth is governed by the principle and precepts of nonviolence?

Is it too late to avoid runaway climate change? It depends on who you listen to. Myles Allen, one of the lead authors of the IPCC reports, says that greenhouse gas emissions are definitely fixable if the fossil fuel industry cleans up its waste. He advises that we should not allow people to sell products that cause climate change. This needs a clear steer from governments, moving from hidden fossil fuel subsidies to carbon taxing and other regulations. However, this doesn't address the carbon that's already in the atmosphere, so perhaps he is too optimistic.

It may well be too late unless we reduce our emissions to stay well below a 2°C temperature rise. 2019 saw a sea change in public attitudes, fed by media coverage from David Attenborough, Greta Thunberg, the school strikes, Extinction Rebellion, and, in the USA, the Sunrise Movement and the Green New Deal. With this growth in awareness there is an opportunity for further radical campaigning to put pressure on political leaders, including nonviolent disruption.

Will short-sightedness be our undoing? We are a long way from the universal sense of responsibility that Sangharakshita recommends. We have been trading off today against tomorrow. To address the present emergency we can extend our horizons to the longer-term future. We can value the future. The basic question for us is, what kind of ancestors do we want to be? Could we be the ancestors who planted trees whose shade we probably won't live to enjoy?

Is there still hope for us? If you're already treating the climate and ecological emergency as an emergency, then yes, there's hope for us. If you're not treating it as an emergency, there's no hope for us!

What do we have hope in? In the Buddhist tradition, there is no creator god that we can petition to intervene on our behalf. The Buddhist equivalent of hope is *sraddha*, confidence or trust. However, *sraddha* is not hope or faith in the ordinary sense. *Sraddha* represents a higher or broader perspective, our connection with vision. It signifies an emotional response to our ideals. In terms of the burning house it represents the father's cry of inspiration.

According to Buddhist psychology *sraddha* provides the bedrock for all positive (skilful) mental states.[3] It is made up of three elements: a deep conviction in what is real, lucidity as to what is of value, and a longing for those qualities that we can attain.

When we're engaged with the potentially divisive and depressing issue of climate and ecological emergency, we need to keep a bigger perspective. What precisely gives Buddhists this bigger perspective? At the highest level, we have *sraddha* in the Three Jewels of the Buddha, the Dharma, and the Sangha. The Buddha jewel symbolizes Gautama the historical Buddha, and the ideal of human Enlightenment, which is present at all times and in all places. The Dharma jewel symbolizes the Buddha's teaching, which is the path to Enlightenment. The Sangha jewel symbolizes the assembly of Enlightened beings down the ages, and more broadly the community of those practising the Dharma.

So when we're talking about hope in a Buddhist context, what we're really talking about is our emotional response to these three ideals. I have *sraddha* that the climate and ecological emergency has the potential to awaken us individually and collectively to Enlightenment. It shows us in the starkest possible terms the consequences of our collective greed, hatred, and delusion. It's as though a giant mirror has been held up to us. Naturally we find looking in this mirror very unpleasant. It goes against the way we want to see ourselves and the way we want others to see us.

We can also have *sraddha* in the Dharma, the Buddha's teaching, perhaps the most famous conceptual formulation of which is dependent origination, that things emerge in dependence upon conditions. An emergency implies *emergence.* We can have *sraddha* that things will arise dependent upon conditions. Sometimes we're not aware of those conditions, like the extreme sensitivity of our climate and civilization to temperature rises, until we perceive the unwelcome consequences. Nevertheless, we can have *sraddha* in the emergent quality of phenomena, and remember that, in an emergency, we can trust in emergence.

We can also have confidence and trust in karma, *how* actions have consequences. Actions based on greed, hatred, and delusion will lead to suffering for ourselves and others, and actions based on generosity, love, and awareness will lead to happiness. As quoted earlier, Martin Luther King said, 'Let us realize that the arc of the moral universe is long, but it bends toward justice.'[4]

To put it into perspective, even if we succeeded in addressing the climate and ecological emergency on its own terms, we would still be faced with the greater problem of what to do with our lives. Where do we find meaning? We would still have to face the problems of impermanence and death. What is born has to die. Everything is impermanent. Even the Buddha had to die. We can be certain of our death; this is an essential aspect of our humanity. The Buddha sought Enlightenment or awakening. He faced up to the problem of birth and death. The climate and ecological emergency is an alarm call for us to wake up from the sleep of unawareness.

Given that actions have consequences depending on the motivations with which they are performed, it's not just what we do but *how* we respond that is crucial. How we respond now will provide the template for future responses. Acceptance, compassion, cooperation, and empathy will lead to different outcomes from aggression, competition, blame, and denial. I suggest that you take your best aspirations and engage in social action based on them. If you have the impulse to benefit society or benefit the world, nurture it and act on it as best you can. Whatever the outcome, doing this will certainly change you, and that is the start of the change you want to see in the world.

In the last chapter we considered nonviolent disruption as a response to the scale of the emergency. It makes a difference *who* is doing the disrupting. A line of meditators creates a compelling image of human potential, in stark contrast to the seriousness of the climate and ecological emergency. Tejopala, whom we met in chapters 5 and 8, organizes interfaith climate and ecological actions. He notices that climate activists, even hard-core atheists, are moved when religious people of any tradition show up and are willing to get arrested. Their presence stimulates faith and represents a bigger perspective, that of a kind of 'moral authority'.

Yogaratna, who appeared in chapter 14, told me about a venerable UK rabbi, Jeffrey Newman. Video footage shows him sitting on the street outside the Bank of England as part of an XR action and being arrested while holding objects sacred to Jews. It's difficult to overstate the impact of this kind of action by a faith leader. Moments before being arrested and dragged away by police, he said, 'If it takes an arrest to try to find ways of helping to galvanise public opinion, then it is *certainly* worth being arrested.'[5]

As Thich Nhat Hanh starkly points out in the opening quotation, if we continue to abuse the earth there is no doubt that our civilization will be destroyed. The goal of Buddhism is to wake up to reality. We can wake up to what is going on around us, what we are losing in terms of quality of life, loss of biodiversity, and loss of wellbeing for future generations. This requires a collective awakening.

The Covid-19 pandemic has revealed the power of solidarity, resilience, equity, and responsibility in our societies. We've seen again that shared risk creates a sense of community. People willingly accepted their part of the collective hardship, benefiting from a greater sense of common purpose and social belonging. Most nations, through the agency of their governments, have prioritized health over economic competitiveness. However, the pandemic has also laid bare long-standing social inequities. As it plays out, it provides an opportunity for a social reset. It also provides an opportunity for a climate and ecological emergency reset.

Could we imagine a world in which there is no fossil fuel pollution, and all energy is renewable? In this world we could end the slaughter of billions of animals annually, and instead grow nutritious, healthy plants. The earth could recover and replenish again the richness that was enjoyed by previous generations.

How can we bring such a world into being? In chapters 7 and 13 I made some suggestions of steps we could take.

Broadly, we need to *de*carbonize our individual lives and collective economics and *re*carbonize the earth through protecting existing carbon sinks and rewilding other areas. Doing this will entail a transition from the Industrial Growth Society to Life-Serving Systems. None of us can act on our own in such a way that we can bring about this transition. Collectively we have a voice and can bring about change.

As part of this transition, what are you personally wanting to offer? Do you buy things that will last and look after them? Could you eat less animal and more plant protein? What are you spending your money and resources on? Could you do without a car and avoid flying? Could you commit to learning about the science behind the climate and ecological emergency and tell the truth about it in whatever forums you participate in? Could you recognize that *we* are the ones who are endangering the lives of future generations?

Are your savings managed in a way that will lead to decarbonization of the economy and recarbonization of the environment? If you are part of a company or institution, do you have a plan to do this? How can it be improved? The climate and ecological emergency requires concerted action that transcends conventional political lines. Do you have a plan for how you can support social change without necessarily taking a party political stance? Will it involve nonviolent disruption? The most important shift in community organizing is from being someone who turns up at an action organized by others, to being someone who asks others. Are you willing to organize actions yourself and ask others to participate in them?

When it comes to society-wide changes it's important to ensure a just transition. The climate and ecological emergency has increased inequalities. How can we bring about decarbonization in an equitable way? Can we support the welfare of those currently working in fossil fuel industries? Can we support the welfare and dignity of People of Colour, the poorest members of society, and indigenous people? Do we need to ensure universal basic services? How can we support girls' education and access to family planning – two changes that combined would produce the greatest impact on climate emissions of any societal changes? How can we care for the soil, introducing regenerative agriculture?

In the field of engineering, we already have the capacity to make radical changes. We can electrify everything that can be electrified, provided that we can clean up battery production. We can green power generation, transport, and industry. We can continue to research and use biofuels. We can introduce efficiencies in building design and construction. We don't currently have the technology to cut around 25 per cent of emissions, so we need to look into carbon capture and even geo-engineering, without relying on these untried technologies for a solution.

To do all of this requires a strong political consensus. Over 125 countries have already committed to carbon neutrality, or Net Zero.[6] This means balancing emissions of carbon dioxide with its removal, often through carbon offsetting or by simply eliminating emissions altogether.

This is a move in the direction of decarbonization, although actual regulations and policies are slow to come. Nevertheless, values are changing, and people are re-evaluating their choices in this area. Politics is also a market, subject to sudden swings in opinion!

For example, what can your country offer in relation to decarbonization and recarbonization? Does it have a plan? Can it be improved? How can we encourage citizen engagement at every level of society?

When it comes to finances, we can transform our notion of value and remember that 'Not everything that counts can be counted.'[7] Western nations have been in the grip of the ideology of free market capitalism since the 1980s. Those who subscribed to this ideology tried to liberate free markets in every aspect of our lives and push against collective action of any kind. But the free market was never going to do anything about the climate and ecological emergency that it created. To give an analogy, once common land has been made barren by overgrazing, companies can move their cattle elsewhere. However, this only works a finite number of times and ignores the impact on locals, who are left to deal with barren common land.

Can we shift from GDP (gross domestic product) as a measure of a nation's wealth to measures of wellbeing? How effectively are people's needs being met, on different levels? Are their physical needs being met, for food, water, shelter, and security? Are their emotional needs being met, for connection, care, empathy, and community? And are their spiritual needs being met, for meaning, value, creativity, freedom, beauty, and wholeness?

Specifically, the fossil fuel industry benefits from the support of governments amounting to a colossal hidden subsidy. Nearly all of what the climate and ecological emergency destroyed is not financially valued.[8] With each tonne of carbon emitted, the harm to society (its contribution to global warming) isn't accounted for or taxed. This results in unreasonably high profits for fossil fuel companies and encourages carbon-emitting activities. Some governments, such as the USA, also directly subsidize the price of fossil fuels to the consumer. The IMF projected these hidden subsidies to be $5.2 trillion in 2017.[9] Rather than subsidizing fossil fuels, can we tax emissions to stimulate decarbonization? Are you willing to campaign on this issue?

Is there any priority to any of these areas: individual choices, a just transition, engineering, politics, and finance? No – to protect the future of life on earth, I suggest that we do everything, and do it now.

My statement turns out to be part cry of anguish, part cry of inspiration. Perhaps we need both, the beauty and the terror. Could we hold both in our hearts? Could we hold the earth, and life itself, as beautiful precisely *because* they are transient? As the famous *Diamond Sutra* says,

> So shall you think of all this fleeting world:
> A star at dawn, a bubble in a stream;
> A flash of lightning in a summer cloud,
> A flickering lamp, a phantom, and a dream.[10]

I hope that these reflections have stimulated your own thoughts, and of course your resolve to act! This is my witness statement. A new world is just waiting to be born. On a quiet day, I can hear it breathing.[11]

Guided reflection: the Great Turning
Go to https://www.windhorsepublications.com/audio-resources/audio-resources-for-tbh/

- Purpose of this activity: to find a broader perspective by imaginatively engaging with a being from seven generations hence.
- Time: 15 minutes.

Appendix A

List of needs

Survival

Air/water/light	Movement	Shelter
Food	Physical safety	Touch
Health	Rest/sleep	

Protection

Order/structure	Security	Trusting
Safety	Stability	

Wellbeing

Balance	Healing	Vitality
Ease	Peace of mind	

To matter

Compassion	Respect	Understanding
Consideration	To be heard and seen	
Empathy	To be valued	

Regeneration

Celebration of life	Leisure	Play
Gratitude	Mourning	

Transcendence

Beauty	Flow	Peace
Communion	Hope	Presence
Faith	Inspiration	Wholeness

Connection

Belonging	Love	Warmth
Closeness	Support	
Harmony	Tenderness	

Freedom

Autonomy	Ease	Spontaneity
Choice	Empowerment	
Control	Self-responsibility	

Honesty

Authenticity	Integrity	Self-connection
Clarity	Openness	Self-expression

Meaning

Challenge	Dignity	Purpose
Consciousness	Growth	Understanding
Contributing to life	Learning	
Creativity	Mastery	

Note: This list is provided only as a tool for study. No list is a substitute for finding your truth using the words that fit your experience. This list builds on Marshall Rosenberg's original needs list (CNVC.org) with categories adapted from Manfred Max-Neef.[1]

Appendix B

List of feelings

Pleasant feelings

Curious

Alert
Amazed
Eager
Energized
Enthusiastic
Excited
Fascinated
Inspired
Interested
Intrigued
Playful

Happy

Confident
Delighted
Encouraged
Grateful
Hopeful
Joyful
Optimistic
Overjoyed
Proud
Relieved

Loving

Affectionate
Appreciative
Compassionate
Friendly
Moved
Nurtured
Open
Sensitive
Tender
Warm

Peaceful

Alive
Calm
Comfortable
Content
Fulfilled
Relaxed
Satisfied
Secure
Strong

Unpleasant feelings

Sad

Ashamed
Depressed
Despairing
Discouraged
Disheartened
Dismayed
Distressed
Guilty
Heavy
Hurt
Unhopeful

Angry

Aggravated
Annoyed
Appalled
Disgusted
Enraged
Furious
Indignant
Infuriated
Irritated
Jealous
Livid
Resentful

Frustrated

Agitated
Bored
Disappointed
Embarrassed
Exasperated
Helpless
Impatient
Tired
Upset

Surprised

Bewildered
Confused
Hesitant
Insecure
Puzzled
Shocked
Torn
Troubled

Afraid

Alarmed
Anxious
Fearful
Frightened
Nervous
Panicky
Scared
Terrified
Worried

Appendix C

Retreat centres

Ecodharma Retreat Centre, Catalunya, Spain offers courses, events, and retreats which support the realization of our human potential and the development of an ecological consciousness honouring our mutual belonging within the web of life – drawing on the Buddhist Dharma and the emerging ecological paradigms of our time. Website: www.ecodharma.com.

Rocky Mountain Ecodharma Retreat Center, Colorado, USA, a home for meditation in nature. The centre, of which David Loy is a director and vice-president, promotes three components of ecodharma: practising in nature, clarifying the ecological implications of Buddhism, and using that understanding to engage in the eco-activism that our situation requires. Website: https://rmerc.org.

Suggestions for further reading

Introduction

Dalai Lama and Franz Alt, *Our Only Home: A Climate Appeal to the World*, Hanover Square Press, Toronto 2020.

J. Stanley, D.R. Loy and G. Dorje (eds), *A Buddhist Response to the Climate Emergency*, Wisdom Publications, Boston 2009.

Chapter 1

A. Leiserowitz, E. Maibach, C. Roser-Renouf, G. Feinberg, and S. Rosenthal, *Global Warming's Six Americas*, Yale University and George Mason University, New Haven, CT, Yale Program on Climate Change Communication 2015, available at https://climatecommunication.yale.edu/visualizations-data/six-americas/. *Global Warming's Six Americas* discovered that there are six distinct groups of people when it comes to attitudes towards the climate/ecological emergency.

Sangharakshita (trans.), *Dhammapada: The Way of Truth*, Windhorse Publications, Birmingham 2001.

Shantigarbha, *I'll Meet You There: A Practical Guide to Empathy, Mindfulness and Communication*, Windhorse Publications, Cambridge 2018.

Subhuti with Sangharakshita, 'Re-imagining the Buddha', in *Seven Papers* (revised 2018), available at https://thebuddhistcentre.com/triratna/seven-papers-subhuti-sangharakshita-direct-download, accessed on 12 March 2021.

Chapter 2

S. Diaz, J. Settele, et al., *Summary for Policymakers of the Global Assessment Report on Biodiversity and Ecosystem Services of the Intergovernmental Science-Policy Platform on Biodiversity and Ecosystem Services 2019*, available at https://www.un.org/sustainabledevelopment/blog/2019/05/nature-decline-unprecedented-report/.

IPCC, 'Summary for policymakers', in *Global Warming of 1.5°C: An IPCC Special Report on the Impacts of Global Warming of 1.5°C above Pre-Industrial Levels and Related Global Greenhouse Gas Emission Pathways, in the Context of Strengthening the Global Response to the Threat of Climate Change, Sustainable Development, and Efforts to Eradicate Poverty*, ed. V. Masson-Delmotte et al., World Meteorological Organization, Geneva 2018.

M.E. Mann, *The New Climate War: The Fight to Take back Our Planet*, Scribe, London 2021.

G. Marshall, *Don't Even Think about It: Why Our Brains Are Wired to Ignore Climate Change*, Bloomsbury, New York 2014.

Chapter 3

S. Kaza and K. Kraft (eds), *Dharma Rain: Sources of Buddhist Environmentalism*, Shambhala Publications, Boston, MA 2000.

P.D. Ryan, *Buddhism and the Natural World: Towards a Meaningful Myth*, Windhorse Publications, Birmingham 1998.

M.E. Tucker and D.R. Williams (eds), *Buddhism and Ecology: The Interconnection of Dharma and Deeds*, Harvard University Center for the Study of World Religions, Cambridge, MA 1997.

Chapter 4

F.H. Cook, *Hua-yen Buddhism: The Jewel Net of Indra*, Pennsylvania State University Press, University Park, PA 2010.

D. Loy, *Ecodharma: Buddhist Teachings for the Ecological Crisis*, Wisdom Publications, Somerville, MA 2018. David Loy addresses the reluctance of Buddhists to engage with climate and environmental activism, which I also address in chapters 4, 5, and 7, where I say that nobody gets to sit this one out.

Sangharakshita, *Living with Awareness: A Guide to the Satipatthana Sutta*, Windhorse Publications, Cambridge 2003.

Subhuti with Sangharakshita, 'Re-imagining the Buddha', in *Seven Papers* (revised 2018), available at https://thebuddhistcentre.com/triratna/seven-papers-subhuti-sangharakshita-direct-download, accessed on 12 March 2021.

Chapter 5

The Karmapa, Ogyen Trinley Dorje, *The Heart Is Noble: Changing the World from the Inside Out*, Shambhala Publications, Boulder, CO 2013.

M.B. Rosenberg, *Nonviolent Communication: A Language of Life*, 3rd ed., Puddledancer, Encinitas, CA 2015.

Chapter 6

S. Diaz, J. Settele, et al., *Summary for Policymakers of the Global Assessment Report on Biodiversity and Ecosystem Services of the Intergovernmental Science-Policy Platform on Biodiversity and Ecosystem Services 2019*, available at https://www.un.org/sustainabledevelopment/blog/2019/05/nature-decline-unprecedented-report/.

R. Layard, *Happiness: Lessons from a New Science*, Penguin, London 2005.

D. Loy, *A Buddhist History of the West*, State University of New York Press, Albany, NY 2002.

M. Max-Neef, *From the Outside Looking In: Experiences in 'Barefoot Economics'*, Zed Books, London 1992. Originally published in 1982 (Dag Hammarskjöld Foundation, Uppsala).

Chapter 7

B. Kato, Y. Tamura, and K. Miyasaka, with revisions by W.E. Soothill, W. Schiffer, and P.P. Del Campana, *The Threefold Lotus Sutra*, Kosei Publishing Co, Tokyo 1990.

W.J. Ripple, C. Wolf, T.M. Newsome, P. Barnard, W.R. Moomaw, and 11,258 scientist signatories from 153 countries, 'World scientists' warning of a climate emergency', *BioScience* 70:1 (January 2020).

Sangharakshita, *The Drama of Cosmic Enlightenment: Parables, Myths and Symbols of the White Lotus Sutra*, Windhorse Publications, Cambridge 1993.

W. Steffen et al., 'Trajectories of the earth system in the Anthropocene', *Proceedings of the National Academy of Sciences* 115 (2018), 8252–9.

S.F.B. Tett, P.A. Stott, M.R. Allen, W.J. Ingram, and J.F.B. Mitchell, 'Causes of twentieth-century temperature change near the earth's surface', *Nature* 399:6736 (1999), 569–72.

Chapter 8

M.C. Hyers, *Zen and the Comic Spirit*, Rider, London 1974.

Chapter 9

Akuppa, *Saving the Earth*, Windhorse Publications, Cambridge 2009.

M.B. Rosenberg, *Nonviolent Communication: A Language of Life*, 3rd ed., Puddledancer, Encinitas, CA 2015, chapter 10: 'Expressing anger fully'.

Sangharakshita, *Know Your Mind: The Psychological Dimension of Ethics in Buddhism*, Windhorse Publications, Birmingham 1998.

Chapter 10

J. Macy and M.Y. Brown, *Coming back to Life: Practices to Reconnect Our Lives, Our World*, New Society Publishers, Gabriola Island, BC 1998.

L. Vaughan-Lee, *Spiritual Ecology: The Cry of the Earth*, 2nd ed., The Golden Sufi Center, Point Reyes, CA 2016.

F. Weller, *The Wild Edge of Sorrow: Rituals of Renewal and the Sacred Work of Grief*, North Atlantic Books, Berkeley, CA 2015.

Chapter 11

Gampopa, *The Jewel Ornament of Liberation: The Wish-Fulfilling Gem of the Noble Teachings*, Snow Lion Publications, Ithaca, NY 1998. The version I've given here mainly follows Vishvapani in his article 'The four reminders', *Madhyamavani*, issue 8, Birmingham 2003, available at http://madhyamavani.fwbo.org/8/reminders.html, accessed on 14 November 2020.

J. Macy and M.Y. Brown, *Coming back to Life: Practices to Reconnect Our Lives, Our World*, New Society Publishers, Gabriola Island, BC 1998.

M.E.P. Seligman, T.A. Steen, N. Park, and C. Peterson, 'Positive psychology progress: empirical validation of interventions', *American Psychologist* 60:5 (2005), 410–21.

Chapter 12

D. Keer, *Dr. Ambedkar: Life and Mission*, 3rd ed., Popular Prakashan, Bombay 1971.

J. Macy and M.Y. Brown, *Coming back to Life: Practices to Reconnect Our Lives, Our World*, New Society Publishers, Gabriola Island, BC 1998.

P.D. Ryan, *Buddhism and the Natural World: Towards a Meaningful Myth*, Windhorse Publications, Birmingham 1998.

Sangharakshita, 'Buddhism, world peace and nuclear war', reprinted in *The Priceless Jewel*, Windhorse Publications, Glasgow 1993, pp.113–36.

Sangharakshita, *Transforming Self and World: Themes from the Sutra of Golden Light*, Windhorse Publications, Birmingham 1995.

M. Stephan and E. Chenoweth, 'Why civil resistance works: the strategic logic of nonviolent conflict', *International Security* 33:1 (summer 2008).

Chapter 13

C. Carson and K. Shepard (eds), *A Call to Conscience: The Landmark Speeches of Martin Luther King, Jr.*, Grand Central Publishing, New York 2001.

M. Max-Neef, *From the Outside Looking In: Experiences in 'Barefoot Economics'*, Zed Books, London 1992. Originally published in 1982 (Dag Hammarskjöld Foundation, Uppsala).

M. Rosenberg, *The Heart of Social Change: How to Make a Difference in Your World*, Puddledancer, Encinitas, CA 2005.

Shantigarbha, *I'll Meet You There: A Practical Guide to Empathy, Mindfulness and Communication*, Windhorse Publications, Cambridge 2018.

Chapter 14

A. Jay et al., 'Overview', in *Impacts, Risks, and Adaptation in the United States: Fourth National Climate Assessment*, D.R. Reidmiller et al., Volume 2, US Global Change Research Program, Washington, DC 2018.

Kutadanta Sutta, DN 5/i, 135.

D. Loy, *Ecodharma: Buddhist Teachings for the Ecological Crisis*, Wisdom Publications, Somerville, MA 2018.

J. Porritt, *Hope in Hell: A Decade to Confront the Climate Emergency*, Simon and Schuster, London 2020.

Sangharakshita, 'Buddhism, world peace and nuclear war', reprinted in *The Priceless Jewel*, Windhorse Publications, Glasgow 1993, pp.113–36.

Sangharakshita, 'Fifteen points for old (and new) Order members', a talk given in 1993, available at https://www.freebuddhistaudio.com/texts/read?num=180&at=text, accessed on 27 December 2020.

Chapter 15

C. Carson and K. Shepard (eds), *A Call to Conscience: The Landmark Speeches of Martin Luther King, Jr.*, Grand Central Publishing, New York 2001.

Dalai Lama and Franz Alt, *Our Only Home: A Climate Appeal to the World*, Hanover Square Press, Toronto 2020.

A.F. Price and W. Mou-Lam (trans.), *The Diamond Sutra and the Sutra of Hui Neng*, Shambhala Publications, Boston, MA 1969.

Sangharakshita, *Know Your Mind: The Psychological Dimension of Ethics in Buddhism*, Windhorse Publications, Birmingham 1998.

Notes

Introduction

1 The Buddha sought an end to samsara, the cycle of death and rebirth, through achieving nirvana (Enlightenment).

2 Dalai Lama and Franz Alt, *Our Only Home: A Climate Appeal to the World*, Hanover Square Press, Toronto 2020.

1. A crisis of empathy

1 *Dhammapada* vv.129–30. See Sangharakshita (trans.), *Dhammapada: The Way of Truth*, Windhorse Publications, Birmingham 2001.

2 A study called *Global Warming's Six Americas* discovered that there are six distinct groups of people when it comes to attitudes towards the climate/ecological emergency. A. Leiserowitz, E. Maibach, C. Roser-Renouf, G. Feinberg, and S. Rosenthal, *Global Warming's Six Americas*, Yale University and George Mason University, New Haven, CT, Yale Program on Climate Change Communication 2015, available at https://climatecommunication.yale.edu/visualizations-data/six-americas/.

3 Subhuti with Sangharakshita, 'Re-imagining the Buddha', in *Seven Papers* (revised 2018), p.17. Available at https://thebuddhistcentre.com/triratna/seven-papers-subhuti-sangharakshita-direct-download.

4 O. Lyons, 'An Iroquois perspective', in *American Indian Environments: Ecological Issues in Native American History*, ed. C. Vecsey and R.W. Venables, Syracuse University Press, New York 1980, pp.171–4.

2. Crisis? What crisis?

1 *Kalama Sutta: To the Kalamas* (AN 3.65), translated from the Pali by Thanissaro Bhikkhu, Access to Insight (BCBS Edition), 30 November 2013, http://www.accesstoinsight.org/tipitaka/an/an03/an03.065.than.html.

2 G. Marshall, *Don't Even Think about It: Why Our Brains Are Wired to Ignore Climate Change*, Bloomsbury, New York 2014, pp.228–9.

3 See M.E. Mann, *The New Climate War: The Fight to Take back Our Planet*, Scribe, London 2021.

4 J. Kotter, *Leading Change*, Harvard Business Review Press, Boston, MA 1996.

5 https://edtechhub.org/.

6 IPCC, 'Summary for policymakers', in *Global Warming of 1.5°C: An IPCC Special Report on the Impacts of Global Warming of 1.5°C above Pre-Industrial Levels and Related Global Greenhouse Gas Emission Pathways, in the Context of Strengthening the Global Response to the Threat of Climate Change, Sustainable Development, and Efforts to Eradicate Poverty*, ed. V. Masson-Delmotte et al., World Meteorological Organization, Geneva 2018.

7 S.F.B. Tett, P.A. Stott, M.R. Allen, W.J. Ingram, and J.F.B. Mitchell, 'Causes of twentieth-century temperature change near the earth's surface', *Nature* 399:6736 (1999), 569–72.

8 The term 'falsifiability' was introduced by the philosopher of science Karl Popper in his book *Logik der Forschung* (1934,

revised and translated into English in 1959 as *The Logic of Scientific Discovery*; revised English edition published in 2002 by Routledge Classics, London).

9 https://climateactiontracker.org/global/temperatures/, accessed on 26 January 2021.

10 World Meteorological Organization, 'Greenhouse gas levels in atmosphere reach new record', press release 22112018, 20 November 2018 (quoting Petteri Taalas), available at https://public.wmo.int/en/media/press-release/greenhouse-gas-levels-atmosphere-reach-new-record, accessed on 28 December 2020.

11 S. Diaz, J. Settele, et al., *Summary for Policymakers of the Global Assessment Report on Biodiversity and Ecosystem Services of the Intergovernmental Science-Policy Platform on Biodiversity and Ecosystem Services 2019*, available at https://www.un.org/sustainabledevelopment/blog/2019/05/nature-decline-unprecedented-report/, accessed on 8 January 2021.

12 *Kalama Sutta: To the Kalamas* (AN 3.65). Elsewhere the Buddha explores how to judge whether a person is wise. In AN 4.192 he says that you can only know if a person is skilful or unskilful by living with them for a long time, watching them carefully. It's through dealing with them through adversity and discussion that you can really tell if they are ethical and wise. In MN 110 he identifies the wise as having conviction, conscience, concern; those who are learned, with aroused persistence, unmuddled mindfulness, and good discernment. See also AN 8.54.

3. Touching the earth

1 *Lalitavistara* or *The Noble Great Vehicle Sūtra, 'The Play in Full'*, p.243. The *Lalitavistara* is a Mahayana account of the early life of the Buddha. The episode of the Buddha-to-be touching the earth and calling her to witness comes in chapter 21: 'Conquering Mara'. Translated from the Tibetan by the Dharmachakra

Translation Committee and published in 2013 by 84000. Tibetan reference: Toh 95, Degé Kangyur, vol. 46 (mdo sde, kha), folios 1b–216b.

2 Sn 3.1.

3 Ibid.

4 Though bear in mind that in another, probably earlier, account Mara speaks 'compassionately' to Siddhartha, urging him to live and make merit – striving is too difficult. Siddhartha replies that he doesn't need merit. He has faith, energy, and wisdom. He declares that he can see Mara's armies of sensual desire, hatred, and the like. He'd rather die than submit! *Padhana Sutta: Exertion* (Sn 3.2), translated from the Pali by Thanissaro Bhikkhu, Access to Insight (BCBS Edition), 30 November 2013, http://www.accesstoinsight.org/tipitaka/kn/snp/snp.3.02.than.html. The later *Lalitavistara* version that I've used here reflects a Mahayana re-emphasis of the importance of compassionate, altruistic activity.

5 *Dhammapada* 14, v.188.

6 The three told here are the Rukkhadhamma Jataka, the Kusanjali Jataka, and the Vyaddha Jataka. C.K. Chapple, 'Animals and environment in the Buddhist birth stories', in *Buddhism and Ecology: The Interconnection of Dharma and Deeds*, ed. Mary Evelyn Tucker and Duncan Ryūken Williams, Harvard University Center for the Study of World Religions, Cambridge, MA 1997.

7 MN 57 and DN 16, 2.8.

8 Nanamoli Bhikku, *The Middle Length Discourses of the Buddha*, Wisdom Publications, Boston, MA 1995, no.31, p.302.

9 From the Buddhist Monastic Code, Vinaya, Pācittiya, Part Two (the living plant chapter), 11.

10 Buddhaghosa, *Visuddhimagga: The Path to Purification.* E.W. Burlingame, *Buddhist Legends*, Harvard University Press, Cambridge, MA 1921, pp.98–9.

11 L. de Silva, 'Early Buddhist attitudes towards nature', in *Dharma Rain: Sources of Buddhist Environmentalism*, ed. Stephanie Kaza and Kenneth Kraft, Shambhala Publications, Boston, MA 2000.
12 The famous *Anapanasati (Mindfulness of Breathing) Sutta*, MN 118.
13 MN 32.
14 'Maha Kassapa Thera: at home in the mountains' (excerpt of Thag 18), translated from the Pali by Andrew Olendzki, Access to Insight (BCBS Edition), 2 November 2013, http://www.accesstoinsight.org/tipitaka/kn/thag/thag.18.00x.olen.html.
15 From the *Mahaparanibbana Sutta*, DN 16, 3.2, 3.55–6, and 4.1.
16 MN 28. I want to acknowledge that for some people sensing the earth element in order to ground yourself may not work. I know that for those of you who live in earthquake-prone regions such as New Zealand or Nepal the earth may have a very different significance. You may not associate it at all with a sense of solidity!
17 SN 47.42.
18 Dhivan Thomas Jones, *This Being, That Becomes: The Buddha's Teaching on Conditionality*, Windhorse Publications, Cambridge 2011.
19 AN 62.

4. Environmental ethics

1 Sangharakshita, *A Moseley Miscellany: Prose and Verse, 1997–2012*, Ibis Publications, Coddington 2015.
2 *Avatamsaka (Flower Ornament) Sutra*, chapter 40.
3 F.H. Cook, *Hua-yen Buddhism: The Jewel Net of Indra*, Pennsylvania State University Press, University Park, PA 2010.
4 S. Kaza, 'Green Buddhism', in *When Worlds Converge: What Science and Religion Tell Us about the Story of the Universe and Our Place in It*, ed. C.N. Matthews, M.E. Tucker, and P. Hefner, Open Court, Chicago 2001, pp.293–309.

5 R.F. Gombrich, *How Buddhism Began: The Conditioned Genesis of the Early Teachings*, 2nd ed., Taylor & Francis Group, London 2011.
6 Sangharakshita, *Dhammapada*.
7 *Adittapariyaya Sutta, The Fire Sermon*, SN 35.28.
8 Sangharakshita, *Know Your Mind: The Psychological Dimension of Ethics in Buddhism*, Windhorse Publications, Birmingham 1998, pp.47–8.
9 Sangharakshita, *Living with Awareness: A Guide to the Satipatthana Sutta*, Windhorse Publications, Cambridge 2003, p.62.
10 I have followed the version given in Asvaghosa, *Buddhacarita, or, Acts of the Buddha*, trans. E.H. Johnston, Motilal Banarsidass, Delhi 1992, Canto 5: 'Flight'. For the more familiar and probably later version placed in the context of a ploughing match, see S. Beal, *Romantic Legends of Sakya Buddha: A Translation of the Chinese Version of the Abhiniskramana Sutra*, Kessinger Publishing Company, Whitefish, MT 2003, p.73.
11 Subhuti with Sangharakshita, 'Re-imagining the Buddha', in *Seven Papers* (revised 2018), p.86. Available at https://thebuddhistcentre.com/triratna/seven-papers-subhuti-sangharakshita-direct-download.
12 D. Loy, *Ecodharma: Buddhist Teachings for the Ecological Crisis*, Wisdom Publications, Somerville, MA 2018.
13 https://soundcloud.com/thebuddhistcentre/sets/the-poetry-interviews-sangharakshita-on-an-apology.
14 Sn 1.8.

5. Compassionate action based on wisdom

1 The Karmapa, Ogyen Trinley Dorje, *The Heart Is Noble: Changing the World from the Inside Out*, Shambhala Publications, Boulder, CO 2013, p.92.
2 J. Cribb, *Surviving the 21st Century: Humanity's Ten Great Challenges and How We Can Overcome Them*, Springer, Basel 2016.

3 Shantigarbha, *I'll Meet You There: A Practical Guide to Empathy, Mindfulness and Communication*, Windhorse Publications, Cambridge 2018.
4 Ibid.
5 https://en.wikipedia.org/wiki/List_of_most-polluted_cities_by_particulate_matter_concentration, accessed on 8 July 2020.
6 Longchenpa, *Kindly Bent to Ease Us*, Part 1: *Mind*, translated from the Tibetan by H.V. Guenther, Dharma Publishing, Berkeley 1975, pp.106ff. and p.119.
7 He read Bill McKibben's article in *Rolling Stone* magazine: B. McKibben, 'Global warming's terrifying new math', *Rolling Stone* (2 August 2012).
8 M.B. Rosenberg, *Nonviolent Communication: A Language of Life*, 3rd ed., Puddledancer, Encinitas, CA 2015, pp.129ff.

6. The five precepts

1 W. Whitman, *Leaves of Grass: 1855 Version*, Ann Arbor Editions LLC, Ann Arbor 2003.
2 Sangharakshita, *Puja: The Triratna Book of Buddhist Devotional Texts*, Windhorse Publications, Cambridge 2008. Sangharakshita emphasized the positive precepts on the basis that we need to know what to cultivate and develop as much as if not more than what we need to avoid.
3 *The National Footprint Accounts, 2011 Edition*, Global Footprint Network, Oakland, CA, available at https://www.footprintnetwork.org, accessed on 30 July 2020.
4 M. Max-Neef, *From the Outside Looking In: Experiences in 'Barefoot Economics'*, Zed Books, London 1992. Originally published in 1982 (Dag Hammarskjöld Foundation, Uppsala).
5 *Kutadanta Sutta*, DN 5/i, 135.
6 B.R. Ambedkar, quoted in J. Demakis, *The Ultimate Book of Quotations*, Lulu, Raleigh, NC 2012, p.415.

7 S. Wynes and K.A. Nicholas, 'The climate mitigation gap: education and government recommendations miss the most effective individual actions', *Environmental Research Letters* 12: 074024 (2017).
8 *Cula-dukkhakkhandha Sutta: The Lesser Mass of Stress*, MN 14.
9 D. Loy, *A Buddhist History of the West*, State University of New York Press, Albany 2002, p.209.
10 R. Layard, *Happiness: Lessons from a New Science*, Penguin, London 2005, p.45.
11 Sangharakshita, in a 1988 talk entitled 'The next twenty years', available at https://www.freebuddhistaudio.com/texts/lecturetexts/170_The_Next_Twenty_Years.pdf.
12 https://soundcloud.com/thebuddhistcentre/sets/the-poetry-interviews-sangharakshita-on-an-apology.
13 https://www.ipcc.ch/sr15/chapter/spm/.
14 Diaz, Settele, et al., *Summary for Policymakers of the Global Assessment Report on Biodiversity and Ecosystem Services of the Intergovernmental Science-Policy Platform on Biodiversity and Ecosystem Services 2019.*
15 Taraloka is a centre of the Triratna Buddhist Community, located in Shropshire, England: www.taraloka.org.uk.
16 See for example the Soto Zen Buddhist Association's 2018 'Statement of recognition and repentance', available at https://www.szba.org/szba-statement-on-recognition-and-repentance/, accessed on 2 August 2020.

7. The burning house

1 B. Kato, Y. Tamura, and K. Miyasaka, with revisions by W.E. Soothill, W. Schiffer, and P.P. Del Campana, *The Threefold Lotus Sutra*, Kosei Publishing Co, Tokyo 1990, p.85.
2 A speech to the EU Parliament in Strasbourg, April 2019.
3 As already mentioned in chapter 4. *Adittapariyaya Sutta: The Fire Sermon*, SN 35.28.

4 This is a joke – I'm just warming up for the next chapter! I'm well aware of the ecological problems associated with electric batteries, and I don't regard electric cars as the solution to the climate and ecological emergency.
5 M. Born, *Physik im Wandel meiner Zeit*, Vieweg, Braunschweig 1966.
6 Tett et al., 'Causes of twentieth-century temperature change'.
7 IPCC, *Climate Change 2001: Synthesis Report. A Contribution of Working Groups I, II, and III to the Third Assessment Report of the Intergovernmental Panel on Climate Change*, ed. R.T. Watson and the Core Writing Team, Cambridge University Press, Cambridge and New York 2001, p.5.
8 This is found in IPCC, 'Summary for policymakers', in *Climate Change 2007: Synthesis Report. Contribution of Working Groups I, II and III to the Fourth Assessment Report of the Intergovernmental Panel on Climate Change*, ed. Core Writing Team, R.K. Pachauri, and A. Reisinger, IPCC, Geneva 2007, p.5.

8. Climate comedy

1 Remarks at an event at the World Economic Forum in Davos, January 2020.
2 Sangharakshita, *Wisdom beyond Words: The Buddhist Vision of Ultimate Reality*, Windhorse Publications, Birmingham 1993, p.267.
3 https://www.buddhistinquiry.org/resources/offerings-analayo/mfcc/.
4 E. Chenoweth and M.J. Stephan, *Why Civil Resistance Works: The Strategic Logic of Nonviolent Conflict*, Columbia University Press, Ohio 2011.
5 A charity funding health and educational projects among some of the poorest communities in India and South Asia: www.karuna.org.
6 http://www.helikos.com/.

7 https://www.maffick.com/.
8 M.C. Hyers, *Zen and the Comic Spirit*, Rider, London 1974.
9 Dawn chorus at Taraloka recorded by and used with permission of Maitridevi: www.taraloka.org.uk.

9. Transforming anger

1 Available at https://tricycle.org/magazine/work-reconnects/, accessed on 26 September 2020.
2 *Visuddhimagga* IX, 23.
3 Sangharakshita, *Peace Is a Fire: A Collection of Writings and Sayings*, compiled by Ananda, originally published by Windhorse Publications, Cambridge 1995, available at https://www.sangharakshita.org/_books/peace-fire.pdf, accessed on 8 October 2020.
4 Sangharakshita, *Know Your Mind*.
5 Thanks to Akuppa for this suggestion.
6 *Dhammapada*, v.5. See Sangharakshita, *Dhammapada*.
7 1969 Montreal Bed-in for Peace, available at https://www.facebook.com/watch/?v=2592170077663084, accessed on 1 February 2021.
8 A. Einstein, 'The real problem is in the hearts of men', interview with Michael Amrine in *New York Times Magazine* (23 June 1946), p.7. His actual words there were, 'Many persons have inquired concerning a recent message of mine that "a new type of thinking is essential if mankind is to survive and move to higher levels".'
9 Rosenberg, *Nonviolent Communication*, chapter 10: 'Expressing anger fully'.
10 Thanks to Dharmachari Saraha for the thought.
11 With thanks to Jim and Jori Manske for the framework: www.radicalcompassion.com.

10. Ecological grief

1 T. McKee, 'The geography of sorrow: Francis Weller on navigating our losses', *Sun Magazine* (October 2015), available at https://www.thesunmagazine.org/issues/478/the-geography-of-sorrow, accessed on 7 October 2020.

2 From 'The definition of love' by the English metaphysical poet Andrew Marvell, in A. Marvell, *The Complete Poems*, Penguin Classics, London 1985, p.49.

3 J. Macy and M.Y. Brown, *Coming back to Life: Practices to Reconnect Our Lives, Our World*, New Society Publishers, Gabriola Island, BC 1998.

4 Ibid.

5 Shantigarbha, *I'll Meet You There*, p.129.

6 Asvaghosa, *Buddhacarita*, Canto 5: 'Flight'.

7 Geshe Wangyal, *The Door of Liberation: Essential Teachings of the Tibetan Buddhist Tradition*, rev. ed., Wisdom Publications, Boston, MA 1995, pp.33–9.

8 Ibid.

9 A. Cunsolo and N.R. Ellis, 'Ecological grief as a mental health response to climate change-related loss', *Nature Climate Change* 8 (2018), 275–81.

10 S. Clayton, C.M. Manning, K. Krygsman, and M. Speiser, *Mental Health and Our Changing Climate: Impacts, Implications, and Guidance*, American Psychological Association and ecoAmerica, Washington, DC 2017.

11 https://www.climatepsychologists.com.

12 Jennifer Welwood, available at https://jenniferwelwood.com/poetry/the-dakini-speaks/, accessed on 18 October 2020.

11. Gratitude

1 W. Berry, *New Collected Poems*, Counterpoint, Berkeley 2012.

2 Kukai, *Major Works*, trans. Y.S. Hakeda, Columbia University Press, Columbia, repr. 1972, pp.51–2.

3 Cicero, 'Pro Plancio' 80, in *Cicero Orationes*, Volume 6, Oxford Classical Texts, Clarendon Press, Oxford 1963.
4 'Count your blessings', in K. Osbeck, *101 Hymn Stories*, Kregel Publishers, Grand Rapids, MI 1982, p.54.
5 Nanamoli Bhikku, *The Middle Length Discourses of the Buddha*, no.31, p.302.
6 Another empirically proven activity is the 'gratitude visit'. Participants were given one week to write and then deliver a letter of gratitude in person to someone who had been especially kind to them but had never been properly thanked. M.E.P. Seligman, T.A. Steen, N. Park, and C. Peterson, 'Positive psychology progress: empirical validation of interventions', *American Psychologist* 60:5 (2005), 410–21.
7 R.A. Emmons and M.E. McCullough, 'Counting blessings versus burdens: an experimental investigation of gratitude and subjective well-being in daily life', *Journal of Personality and Social Psychology* 84:2 (2003), 377–89.
8 S.B. Algoe, S.L. Gable, and N.C. Maisel, 'It's the little things: everyday gratitude as a booster shot for romantic relationships', *Personal Relationships* 17 (2010), 217–33.
9 Y. Chang, Y. Lin, and L.H. Chen, 'Pay it forward: gratitude in social networks', *Journal of Happiness Studies* 13 (2012), 761–81.
10 Gampopa, *The Jewel Ornament of Liberation: The Wish-Fulfilling Gem of the Noble Teachings*, Snow Lion Publications, Ithaca, NY 1998.
11 *Chiggala Sutta: The Hole* (SN 56.48), translated from the Pali by Thanissaro Bhikkhu, Access to Insight (BCBS Edition), 1 July 2010, available at http://www.accesstoinsight.org/tipitaka/sn/sn56/sn56.048.than.html.
12 The version of the practice I've given here mainly follows Vishvapani in his article 'The four reminders', *Madhyamavani*, issue 8, Birmingham 2003, available at http://madhyamavani.fwbo.org/8/reminders.html, accessed on 2 February 2021.

13 Macy and Brown, *Coming back to Life*.

12. Nonviolent social change

1 From his speech to the All-India Depressed Classes Conference at Nagpur, 18–19 July 1942. D. Keer, *Dr. Ambedkar: Life and Mission*, 3rd ed., Popular Prakashan, Bombay 1971, p.351.
2 Macy and Brown, *Coming back to Life*.
3 S. Kvaløy, 'Complexity and time: breaking the pyramid's reign', in *Wisdom in the Open Air: The Norwegian Roots of Deep Ecology*, ed. P. Reed and D. Rothenberg, University of Minnesota Press, Minneapolis 1993, chapter 4.
4 Macy and Brown, *Coming back to Life*, p.17.
5 The story of the Buddha mediating between Shakyans and Koliyans is traditionally associated with the teaching given in the *Attadanda Sutta* in the *Sutta Nipata* 935–54 ('Fear results from resorting to violence – just look at how people quarrel and fight!'). Found in G.P. Malalasekera, *Dictionary of Pali Proper Names*, Luzac, London 1960, repr. Asian Educational Services, New Delhi 2003, available at www.metta.lk/pali-utils/Pali-Proper-Names/rohini.htm, accessed on 14 December 2020. It is also associated with the fifth-century *Commentary to the Dhammapada*, found in *Buddhist Stories from the Dhammapada Commentary*, Part 2, translated from the Pali by E.W. Burlingame, selected and revised by Bhikkhu Khantipalo, *Wheel* 324, Buddhist Publication Society, Kandy 1985. There the story is associated with *Dhammapada* vv.197–9 ('Happy indeed we live, friendly amid the haters. Among men who hate we dwell free from hate'). The Rohini incident is retold and discussed in 'Buddhism, world peace and nuclear war' by Sangharakshita, reprinted in *The Priceless Jewel*, Windhorse Publications, Glasgow 1993, available at http://www.sangharakshita.org/_books/The Priceless Jewel.pdf, accessed on 14 December 2020.

6 There are two Ecodharma retreat centres: one in Spain, www.ecodharma.com, and one in Colorado, https://rmerc.org. See the list of retreat centres in Appendix C. The Ecosattva initiative is from Buddhafield: https://www.buddhafield.com.
7 *Sutta Nipata* 1.7, *An Outcaste.*
8 P.D. Ryan, *Buddhism and the Natural World: Towards a Meaningful Myth*, Windhorse Publications, Birmingham 1998, p.42.
9 N.A. Nikam and R. McKeon, *The Edicts of Asoka*, University of Chicago Press, Chicago 1959, p.55.
10 B.R. Ambedkar, *Writing and Speeches: A Ready Reference Manual of 17 Volumes*, B.R. Publishing Corporation, New Delhi 2007.
11 From an address to All India Radio, 1954.
12 *Transforming Self and World: Themes from the Sutra of Golden Light*, Windhorse Publications, Birmingham 1995, p.14.
13 M. Stephan and E. Chenoweth, 'Why civil resistance works: the strategic logic of nonviolent conflict', *International Security* 33:1 (summer 2008).

13. Be the change

1 M.K. Gandhi, *The Collected Works of Mahatma Gandhi*, Volume 12, *April 1913 to December 1914*, The Publications Division, Ministry of Information and Broadcasting, Government of India 1964, p.158.
2 A. Lorrance, 'The love project', in *Developing Priorities and a Style: Selected Readings in Education for Teachers and Parents*, ed. Richard Dean Kellough, MSS Information Corporation, New York 1974, p.85.
3 M. Rosenberg, *The Heart of Social Change: How to Make a Difference in Your World*, Puddledancer, Encinitas, CA 2005, pp.5–6.
4 From 'Where do we go from here?', an address delivered at the 11th annual SCLC Convention, 16 August 1967, Atlanta, GA, in *A Call to Conscience: The Landmark Speeches of Martin Luther King,*

Jr., ed. C. Carson and K. Shepard, Grand Central Publishing, New York 2001.

5 Marshall had various ways of presenting this process, none of which I found easy to learn or share with others. The process here is based on 'Social change as spiritual practice', developed by NVC trainers Charles Jones (www.mobiusleadership.com/practitioners/charles-jones) and Gina Lawrie (www.ginalawrie.co.uk). Over the years I've gradually adapted and modified it to suit the groups I've been working with.

6 For my workshops visit www.SeedofPeace.org.

7 Max-Neef, *From the Outside Looking In*.

14. Climate justice and nonviolent disruption

1 Sangharakshita, 'Buddhism, world peace and nuclear war', pp.113–36.

2 Sangharakshita, 'The Buddha by Trevor Ling, Part 2', 1976, p.2, available at https://www.freebuddhistaudio.com/texts/read?num=SEM054P2, accessed on 31 December 2020.

3 *Kutadanta Sutta*, DN 5/i, 135.

4 *Cakkavatti Sihanada Sutta*, DN 26/iii, 61.

5 D.R. Reidmiller et al. (ed.), *Impacts, Risks, and Adaptation in the United States: Fourth National Climate Assessment*, Volume 2, US Global Change Research Program, Washington, DC 2018, pp. 33–71.

6 Potsdam Institute for Climate Impact Research and Climate Analytics, *Turn down the Heat: Climate Extremes, Regional Impacts, and the Case for Resilience*, World Bank, Washington, DC 2013, available at https://openknowledge.worldbank.org/handle/10986/14000, accessed on 1 January 2021.

7 Greta Thunberg to world leaders at the 2019 UN climate action summit in New York.

8 IPCC, *Climate Change 2001: Synthesis Report*, p.5.

9 https://www.stopecocide.earth/.

10 Max-Neef, *From the Outside Looking In*.

11 António Guterres delivered his speech on climate change and his vision for the 2019 climate change summit on 10 September 2018, available at https://www.un.org/sg/en/content/sg/statement/2018-09-10/secretary-generals-remarks-climate-change-delivered, accessed on 2 February 2021.

12 J. Porritt, *Hope in Hell: A Decade to Confront the Climate Emergency*, Simon and Schuster, London 2020, p.8.

13 Loy, *Ecodharma*.

14 Private conversation with the author, January 2021.

15 *The Buddhist Monastic Code*, Volume 1, *The Patimokkha Rules*, translated and explained by Thanissaro Bhikkhu (Geoffrey DeGraff), 3rd ed. rev., Metta Forest Monastery, Valley Center, CA 2013, p.68.

16 Thanks to Rowan Tilly for articulating these to me.

17 Sangharakshita, 'Fifteen Points for Old (and New) Order Members', a talk given in 1993, transcript available at https://www.freebuddhistaudio.com/texts/read?num=180&at=text, accessed on 27 December 2020.

15. Final thoughts: the beauty and the terror

1 T.N. Hanh, *The Art of Power*, HarperCollins, New York 2008.

2 Sangharakshita, 'Buddhism, world peace and nuclear war'.

3 Sangharakshita, *Know Your Mind*, p.125.

4 From 'Where do we go from here?', an address delivered at the 11th annual SCLC Convention, 16 August 1967, Atlanta, GA, in Carson and Shepard (ed.), *A Call to Conscience*.

5 https://twitter.com/damiengayle/status/1183708840908509184?s=20.

6 https://www.carbontrust.com/what-we-do/net-zero.

7 W.B. Cameron, *Informal Sociology: A Casual Introduction to Sociological Thinking*, Random House, New York 1963, p.13.

8 P. Dasgupta, *The Economics of Biodiversity: The Dasgupta Review*, HM Treasury, London 2021, available at https://www.gov.uk/government/publications/final-report-the-economics-of-biodiversity-the-dasgupta-review, accessed on 5 February 2021.

9 D. Coady, I. Parry, N.-P. Le, and B. Shang, 'Global fossil fuel subsidies remain large: an update based on country-level estimates', IMF Working Paper 19/89 2019, available at https://www.imf.org/en/Publications/WP/Issues/2019/05/02/Global-Fossil-Fuel-Subsidies-Remain-Large-An-Update-Based-on-Country-Level-Estimates-46509, accessed on 12 March 2021.

10 Translated by Dr Kenneth Saunders, quoted in A.F. Price and W. Mou-Lam (trans.), *The Diamond Sutra and the Sutra of Hui Neng*, Shambhala, Boston, MA 1969, p.74.

11 With thanks to Arundhati Roy for the thought.

Appendix A

1 M. Max-Neef et al., 'Human scale development: an option for the future', *Development Dialogue* 1 (1989), 5–80.

Index

Introductory Note

References such as '178–9' indicate (not necessarily continuous) discussion of a topic across a range of pages. Wherever possible in the case of topics with many references, these have either been divided into sub-topics or only the most significant discussions of the topic are listed. Because the entire work is about 'climate', the use of this term (and certain others which occur constantly throughout the book) as an entry point has been minimized. Information will be found under the corresponding detailed topics.

Index

WINDHORSE PUBLICATIONS

Windhorse Publications is a Buddhist charitable company based in the United Kingdom. We place great emphasis on producing books of high quality that are accessible and relevant to those interested in Buddhism at whatever level. We are the main publisher of the works of Sangharakshita, the founder of the Triratna Buddhist Order and Community. Our books draw on the whole range of the Buddhist tradition, including translations of traditional texts, commentaries, books that make links with contemporary culture and ways of life, biographies of Buddhists, and works on meditation.

As a not-for-profit enterprise, we ensure that all surplus income is invested in new books and improved production methods, to better communicate Buddhism in the 21st century. We welcome donations to help us continue our work – to find out more, go to windhorsepublications.com.

The Windhorse is a mythical animal that flies over the earth carrying on its back three precious jewels, bringing these invaluable gifts to all humanity: the Buddha (the 'awakened one'), his teaching, and the community of all his followers.

Windhorse Publications
38 Newmarket Road
Cambridge CB5 8DT
info@windhorsepublications.com

Consortium Book Sales & Distribution
210 American Drive
Jackson TN 38301
USA

Windhorse Books
PO Box 574
Newtown NSW 2042
Australia

THE TRIRATNA BUDDHIST COMMUNITY

Windhorse Publications is a part of the Triratna Buddhist Community, an international movement with centres in Europe, India, North and South America and Australasia. At these centres, members of the Triratna Buddhist Order offer classes in meditation and Buddhism. Activities of the Triratna Community also include retreat centres, residential spiritual communities, ethical Right Livelihood businesses, and the Karuna Trust, a United Kingdom fundraising charity that supports social welfare projects in the slums and villages of India.

Through these and other activities, Triratna is developing a unique approach to Buddhism, not simply as a philosophy and a set of techniques, but as a creatively directed way of life for all people living in the conditions of the modern world.

If you would like more information about Triratna please visit thebuddhistcentre.com or write to:

London Buddhist Centre
51 Roman Road
London E2 0HU
United Kingdom

Aryaloka
14 Heartwood Circle
Newmarket NH 03857
USA

Sydney Buddhist Centre
24 Enmore Road
Sydney NSW 2042
Australia